AF588223

Advances in Experimental Medicine and Biology

Neuroscience and Respiration

Volume 944

More information about this series at http://www.springer.com/series/13457

Mieczyslaw Pokorski
Editor

Respiratory Treatment and Prevention

Editor
Mieczyslaw Pokorski
Public Higher Medical Professional School in Opole
Institute of Nursing
Opole, Poland

ISSN 0065-2598 ISSN 2214-8019 (electronic)
Advances in Experimental Medicine and Biology
ISBN 978-3-319-44487-1 ISBN 978-3-319-44488-8 (eBook)
DOI 10.1007/978-3-319-44488-8

Library of Congress Control Number: 2016958561

Printed on acid-free paper

This Springer imprint is published by Springer Nature
The registered company is Springer International Publishing AG Switzerland

Preface

The book series *Neuroscience and Respiration* presents contributions by expert researchers and clinicians in the field of pulmonary disorders. The chapters provide timely overviews of contentious issues or recent advances in the diagnosis, classification, and treatment of the entire range of pulmonary disorders, both acute and chronic. The texts are thought as a merger of basic and clinical research dealing with respiratory medicine, neural and chemical regulation of respiration, and the interactive relationship between respiration and other neurobiological systems such as cardiovascular function or the mind-to-body connection. The authors focus on the leading-edge therapeutic concepts, methodologies, and innovative treatments. Pharmacotherapy is always in the focus of respiratory research. The action and pharmacology of existing drugs and the development and evaluation of new agents are the heady area of research. Practical, data-driven options to manage patients will be considered. New research is presented regarding older drugs, performed from a modern perspective or from a different pharmacotherapeutic angle. The introduction of new drugs and treatment approaches in both adults and children also is discussed.

Lung ventilation is ultimately driven by the brain. However, neuropsychological aspects of respiratory disorders are still mostly a matter of conjecture. After decades of misunderstanding and neglect, emotions have been rediscovered as a powerful modifier or even the probable cause of various somatic disorders. Today, the link between stress and respiratory health is undeniable. Scientists accept a powerful psychological connection that can directly affect our quality of life and health span. Psychological approaches, by decreasing stress, can play a major role in the development and therapy of respiratory diseases.

Neuromolecular aspects relating to gene polymorphism and epigenesis, involving both heritable changes in the nucleotide sequence and functionally relevant changes to the genome that do not involve a change in the nucleotide sequence, leading to respiratory disorders will also be tackled. Clinical advances stemming from molecular and biochemical research are but possible if the research findings are translated into diagnostic tools, therapeutic procedures, and education, effectively reaching physicians and patients. All that cannot be achieved without a multidisciplinary, collaborative, bench-to-bedside approach involving both researchers and clinicians.

The societal and economic burden of respiratory ailments has been on the rise worldwide leading to disabilities and shortening of life span. COPD alone causes more than three million deaths globally each year. Concerted efforts are required to improve this situation, and part of those efforts are gaining insights into the underlying mechanisms of disease and staying abreast with the latest developments in diagnosis and treatment regimens. It is hoped that the books published in this series will assume a leading role in the field of respiratory medicine and research and will become a source of reference and inspiration for future research ideas.

I would like to express my deep gratitude to Mr. Martijn Roelandse and Ms. Tanja Koppejan from Springer's Life Sciences Department for their genuine interest in making this scientific endeavor come through and in the expert management of the production of this novel book series.

Opole, Poland Mieczyslaw Pokorski

Contents

Osteoprotegerin/sRANKL Signaling System in Pulmonary Sarcoidosis: A Bronchoalveolar Lavage Study 1
W. Naumnik, B. Naumnik, W. Niklińska, M. Ossolińska, and E. Chyczewska

Ambient PM2.5 Exposure and Mortality Due to Lung Cancer and Cardiopulmonary Diseases in Polish Cities 9
Artur J. Badyda, James Grellier, and Piotr Dąbrowiecki

Non-Tuberculous Mycobacteria: Classification, Diagnostics, and Therapy 19
I. Porvaznik, I. Solovič, and J. Mokrý

Diagnosis of Invasive Pulmonary Aspergillosis 27
E. Swoboda-Kopeć, M. Sikora, K. Piskorska, M. Gołaś, I. Netsvyetayeva, D. Przybyłowska, and E. Mierzwińska-Nastalska

Rarity of Mixed Species Malaria with *Plasmodium falciparum* and *Plasmodium malariae* in Travelers to Saarland in Germany 35
Josef Yayan and Kurt Rasche

Antibiotics Modulate the Ability of Neutrophils to Release Neutrophil Extracellular Traps 47
Aneta Manda-Handzlik, W. Bystrzycka, S. Sieczkowska, U. Demkow, and O. Ciepiela

Sensitivity of Next-Generation Sequencing Metagenomic Analysis for Detection of RNA and DNA Viruses in Cerebrospinal Fluid: The Confounding Effect of Background Contamination 53
Iwona Bukowska-Ośko, Karol Perlejewski, Shota Nakamura, Daisuke Motooka, Tomasz Stokowy, Joanna Kosińska, Marta Popiel, Rafał Płoski, Andrzej Horban, Dariusz Lipowski, Kamila Caraballo Cortés, Agnieszka Pawełczyk, Urszula Demkow, Adam Stępień, Marek Radkowski, and Tomasz Laskus

Mandibular Advancement Appliance for Obstructive Sleep Apnea Treatment 63
J. Kostrzewa-Janicka, P. Śliwiński, M. Wojda, D. Rolski, and E. Mierzwińska-Nastalska

Acute Response to Cigarette Smoking Assessed in Exhaled Breath Condensate in Patients with Chronic Obstructive Pulmonary Disease and Healthy Smokers 73
M. Maskey-Warzęchowska, P. Nejman-Gryz, K. Osinka, P. Lis, K. Malesa, K. Górska, and R. Krenke

Cigarette Smoking and Respiratory System Diseases in Adolescents 81
Agnieszka Saracen

Erratum E1

Index 87

The original version of this book has been revised. An erratum to this book can be found at DOI 10.1007/978-3-319-44488-8_11

Advs Exp. Medicine, Biology - Neuroscience and Respiration (2017) 27: 1–7
DOI 10.1007/5584_2016_44

Published online: 12 July 2016

Osteoprotegerin/sRANKL Signaling System in Pulmonary Sarcoidosis: A Bronchoalveolar Lavage Study

W. Naumnik, B. Naumnik, W. Niklińska, M. Ossolińska, and E. Chyczewska

Abstract

Osteoprotegerin (OPG), a soluble tumor necrosis factor receptor family molecule, protects endothelial cells from apoptosis *in vitro* and promotes neovascularization *in vivo*. Angiogenesis may be crucial for the course and outcome of sarcoidosis. In this study, we evaluated the clinical usefulness of OPG and its ligand, a soluble receptor activator of nuclear factor-kappaB (sRANKL), in bronchoalveolar lavage fluid (BALF) in patients with sarcoidosis (BBS, Besniera-Boeck-Schaumann disease). We studied 22 BBS patients and 15 healthy volunteers as a control group. The levels of OPG, sRANKL, and interleukin-18 (IL-18) were measured by the Elisa method. The BALF levels of sRANKL and IL-18 were higher in the BBS patients compared with controls [sRANKL: 2.12 (0.82–10.23) *vs.* 1.12 (0.79–4.39) pmol/l, $p = 0.03$; IL-18: 34.29 (12.50–133.70) *vs.* 13.05 (12.43–25.88) pg/ml, $p = 0.001$]. There were no significant differences between the concentration of OPG in the BBS patients and healthy controls [0.22 (0.14–0.81) *vs.* 0.23 (0.14–0.75) pmol/l]. In the BBS patients we found correlations between sRANKL and IL-18 in BALF ($r = 0.742$, $p = 0.0001$) and between OPG and lung diffusing capacity for carbon monoxide (DLCO) ($r = -0.528$, $p = 0.029$). Receiver-operating characteristic (ROC) curve was applied to find the cut-off for the BALF level of sRANKL (BBS vs. healthy: 1.32 pmol/l).

The original version of this chapter has been revised. An erratum to this chapter can be found at DOI 10.1007/978-3-319-44488-8_11

W. Naumnik (✉)
First Department of Lung Diseases, Medical University of Bialystok, 14 Zurawia St, 15-540 Bialystok, Poland

Department of Clinical Molecular Biology, Medical University of Bialystok, Bialystok, Poland
e-mail: wojciechnaumnik@gmail.com

B. Naumnik
First Department of Nephrology and Transplantation with Dialysis Unit, Medical University of Bialystok, Bialystok, Poland

W. Niklińska
Department of Histology and Embryology, Medical University of Bialystok, Bialystok, Poland

M. Ossolińska and E. Chyczewska
First Department of Lung Diseases, Medical University of Bialystok, 14 Zurawia St, 15-540 Bialystok, Poland

We conclude that OPG and sRANKL may have usefulness in clinical evaluation of BBS patients.

Keywords
Bronchoalveolar lavage fluid • Lungs • Osteoprotegerin • Soluble receptor activator of nuclear factor-kappaB ligand • Sarcoidosis

1 Introduction

Sarcoidosis is a systemic disease of unknown origin that is characterized by the formation of immune granulomas in various organs, mainly the lungs and the lymphatic system (Valeyre et al. 2014). The imbalance that shifts the Th1/Th2 equilibrium toward Th1 immunity and angiogenesis have been suggested to play a pivotal role in the immunopathogenesis of sarcoidosis (Crowley et al. 2011). There also is an association between interleukin-18 (IL-18) and activity of pulmonary sarcoidosis (Liu et al. 2010). Recent theories implicate an exaggerated immune response to antigens that remain unidentified as of yet (Chen and Moller 2008).

Mortality in patients with sarcoidosis is higher than that of the general population, due mainly to pulmonary fibrosis (Valeyre et al. 2014). Recently, several studies have revealed that the osteoprotegerin/receptor activator for nuclear factor kB ligand (OPG/RANKL) system, traditionally implicated in bone remodelling, relates with the prediction of mortality in humans (Reinhard et al. 2010; Semb et al. 2009). OPG, also known as osteoclastogenesis inhibitory factor, is expressed in many tissues such as the heart, kidney, liver, lung, and bronchi (Sandberg et al. 2006). OPG is a key regulator in vascular diseases and is functionally involved with angiogenesis and the regulation of a cell phenotype (Cross et al. 2006). Moreover, OPG protects endothelial cells from apoptosis *in vitro* and promotes neovascularization *in vivo* (Goswami and Sharma-Walia 2015). OPG acts as a decoy receptor for RANKL, thereby interfering with RANKL-binding to its cell-surface receptor RANK (Sandberg et al. 2006). The RANK-RANKL interaction triggers vascular permeability, cytokine release, monocyte transmigration, and monocyte matrix metalloproteinase activity (Collin-Osdoby 2004). RANKL is expressed in activated T-lymphocytes, lymph nodes, lungs, lower respiratory tract, and spleen (Boyce and Xing 2008.).

The features of the OPG/RANKL system above outlined, in particular its role in angiogensis-related phenomena, make it a potentially influential player in the pathomechanisms of sarcoidosis, i.e., Besniera-Boeck-Schaumann (BBS) disease, the issue that has not yet been scientifically pursued. Therefore, in the present study we seek to determine the levels of OPG, sRANKL, and IL-18 in bronchoalveolar lavage fluid (BALF) in sarcoidosis patients, as compared with healthy control subjects, in an attempt to evaluate the possible predictive role of the OPG/RANKL system in the course of BBS.

2 Methods

The study protocol was approved by a local Ethics Committee and written consent was obtained from all study participants. The study was performed in conformity with the Declaration of Helsinki for Human Experimentation of the World Medical Association and was executed at the Department of Lung Diseases of Bialystok Medical University in Poland during the period of 2010–2014.

2.1 Patients

The study group consisted of 22 patients (18 men; mean age 46.5 ± 9 years) suffering from sarcoidosis diagnosed according to the clinical and pathological guidelines of Costabel and Hunninghake (1999). The disease was at the second stage, with bilateral hilar lymphadenopathy and pulmonary infiltrations, and was histologically confirmed. The control group consisted of 15 healthy volunteers (13 men, mean age 49.7 years) without any inflammatory conditions. The following functional tests were performed: spirometry, plethysmography, and diffusing lung capacity for carbon monoxide (DLCO) according to the guidelines of the American Thoracic Society (1995). Basic diagnostic biochemical work-up, gasometry, and ECG were conducted in each patient. In addition, bronchofiberoscopy (Pentax FB 18 V, Pentax Corporation, Tokyo, Japan) with BALF collection were performed in all patients and control subjects. We used premedication with intramuscular atropine and midazolam, and subsequent local anaesthesia with lidocaine before bronchoscopy. The bronchofiberoscopy was inserted and wedged in the middle right lobe, and three 50 ml aliquots of sterile saline solution, warmed to 37 °C, were instilled and aspirated from the subsegmental bronchus. The recovered fluid was filtered through sterile gauze and centrifuged at 1500 rpm for 15 min at 4 °C. Supernatant was stored at −70 °C until use. Samples of BALF were assayed for total and differential cell counts, including CD4+, and CD8+ lymphocytes, using fluorescence-based flow cytometry (Becton Dickinson; Mountain View, CA). Cell differentials were made in smears stained with the Grünwald-Giemza method by counting at least 400 cells under a light microscope (magnification 1 k). The number of CD4 and CD8 cells was counted a percentage of positively stained cells. The cell suspension of BALF was incubated with fluorescein isothiocyanate-labeled anti-CD8 antibody and phycoerythrin-labeled anti-CD4 antibody (Becton Dickinson) for 20 min, then washed with PBS twice.

2.2 Concentrations of OPG, sRANKL, and IL-18 in BALF

We measured the level of OPG (Biomedica Medizinprodukte; Vienna, Austria), sRANK (BioVendor – Laboratorní medicína a.s.; Brno, Czech Republic), and IL-18 (MBL International Corp.; Woburn, MA) in BALF supernatant using commercially available enzyme-linked immunosorbent assay (ELISA) kits. The minimum detectable levels of OPG, sRANKL and IL-18 were 0.14 pmol/l, 0.1 pmol/l, and 1.5 pg/ml, respectively.

2.3 Statistical Analysis

Data distribution was analyzed with the Shapiro-Wilk test. Normally distributed data were presented as means ± SE. Data with skewed distribution were presented as medians and minimum-maximum ranges. Differences between the mean values were assessed using a *t*-test for normally distributed data and the Mann-Whitney U or Wilcoxon tests for skewed distribution. Correlations between variables were investigated with Spearman's rank test. Receiver-operating characteristic (ROC) curves were used to find out the cut-off levels for sRANKL. A p-value < 0.05 defined statistically significant differences. Statistical elaboration was performed using Statistica 12.0 software (StatSoft Inc., Tulsa, CA).

3 Results

There were no significant differences in age or gender between the patients and healthy subjects. With respect to pulmonary function, BBS patients had a significantly reduced vital capacity (VC) – 86.2 ± 15.2 *vs.* 97.2 ± 3.1 %pred., p = 0.02 and lung diffusion for carbon monoxide (DLCO) – 81.2 ± 25 *vs.* 91.4 ± 11.2 % pred., p = 0.01, compared with healthy subjects.

BALF analysis revealed that the BBS patients had a higher percentage of lymphocytes and a lower percentage of macrophages: lymphocytes – 41.3 ± 20.1 *vs.* 17.1 ± 7 %, p = 0.001 and macrophages – 52.3 ± 19.2 *vs.* 81.2 ± 15 %, p = 0.003. The counts of neutrophils and eosinophils were inappreciably different between the patients and controls. The percentage of CD4 + was higher in BALF of BBS patients than that in controls – 45.1 ± 15.2 *vs.* 8.3 ± 0.2 %, p = 0.001. There was an inappreciable difference in CD8+ between the patients and healthy subjects (16.8 ± 3.2 *vs.* 18.1 ± 4.2 %, p = 0.121). The BALF recovery rate was similar in the BBS and control groups.

BALF levels of OPG, sRANKL, and IL-18 are shown in Fig. 1. The OPG concentration was similar in the BBS patients and healthy subjects [OPG: 0.22 (0.14–0.84) *vs.* 0.23 (0.14–0.75) pmol/l, p = 0.744] (Fig. 1a). The sRANKL and IL-18 concentrations were significantly greater in the BBS patients compared with healthy subjects [sRANKL: 2.12 (0.82–10.23) *vs.* 1.12 (0.79–4.39) pmol/l, p = 0.013; IL-18: 34.29 (12.5–133.70) *vs.* 13.05 (12.43–25.88) pg/ml, p = 0.001] (Fig. 1b, c).

There were significant correlations between sRANKL and IL-18 (r = 0.742, p = 0.0001) (Fig. 2) and OPG and DLCO (r = –0.528, p = 0.029) (Fig. 3) in BALF of BBS patients. Moreover, sRANKL correlated positively with the percentage of lymphocytes (r = 0.60, p = 0.003) and negatively with the percentage of macrophages (r = –0.58, p = 0.005) in these patients. The correlations above outlined were absent in the healthy group.

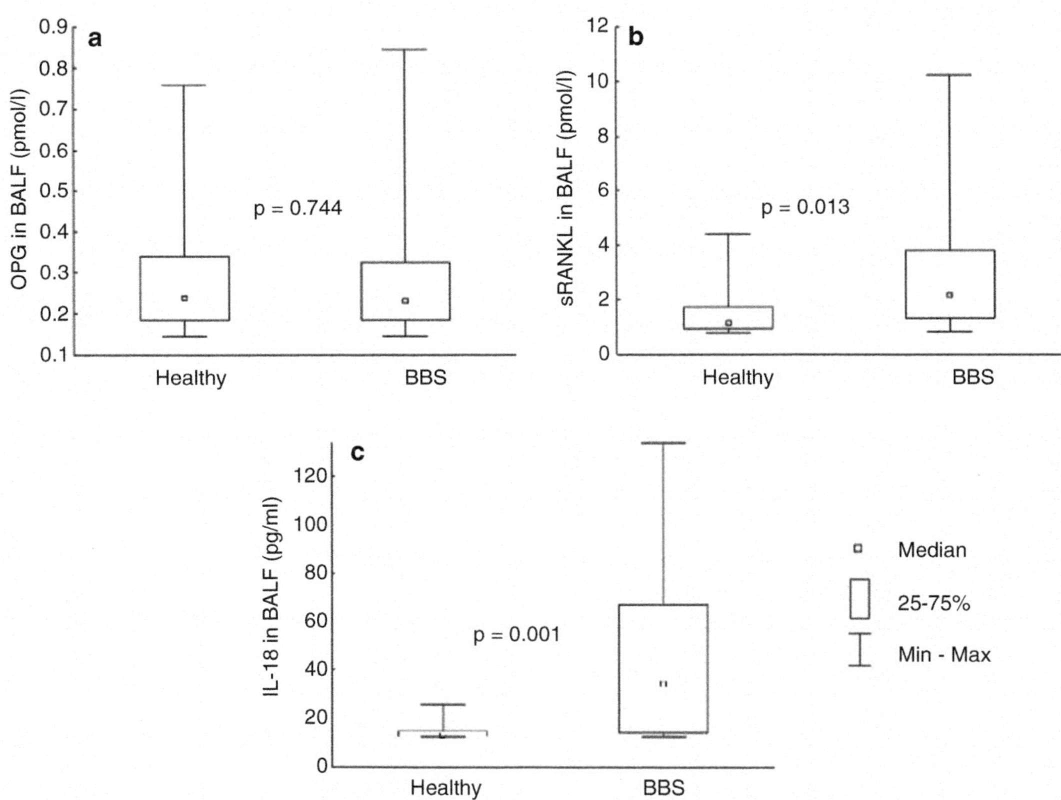

Fig. 1 (**a**) Concentration of osteoprotegerin (OPG); (**b**) Soluble receptor activator of nuclear factor-kappaB (sRANKL); and (**c**) interleukin-18 (IL-18) in bronchoalveolar lavage fluid (BALF) of sarcoidosis (BBS – Besniera-Boeck-Schaumann disease) patients compared with those in healthy subjects

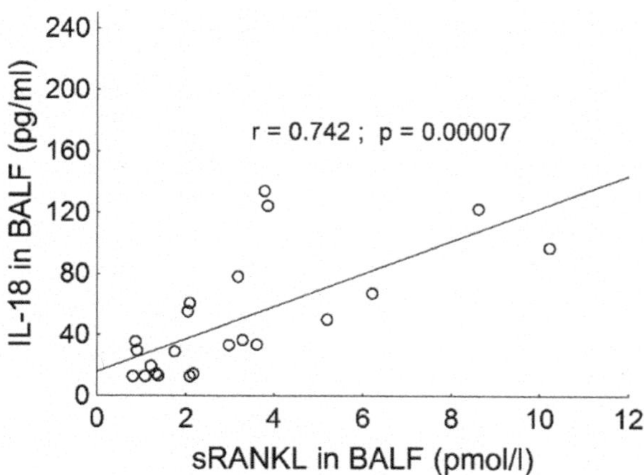

Fig. 2 Correlation between interleukin-18 (IL-18) and soluble receptor activator of nuclear factor-kappaB (*sRANKL*) in bronchoalveolar lavage fluid (*BALF*) of sarcoidosis (*BBS* – Besniera-Boeck-Schaumann disease) patients

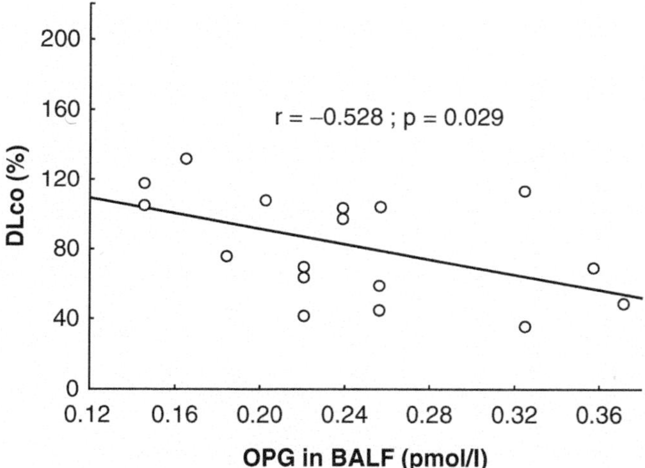

Fig. 3 Correlation between lung diffusing capacity for carbon monoxide (*DLCO*) and osteoprotegerin (*OPG*) in bronchoalveolar lavage fluid (*BALF*) of sarcoidosis (*BBS* – Besniera-Boeck-Schaumann disease) patients (Sperman's rank correlation on median values)

The cut-off value of sRANKL for optimal diagnostic efficiency in distinguishing between patients with sarcoidosis and healthy subjects was 1.32 pmol/l as determined by ROC curve analysis, with AUC 0.67 (Fig. 4).

4 Discussion

The cause of sarcoidosis is still unknown and diagnostics can be difficult and delayed because of diverse, nonspecific, unusual, or initially misleading presentations (Valeyre et al. 2014). Several studies have revealed that angiogenesis and microvascular changes are closely linked with BBS (Zielonka et al. 2007). The present findings are consistent with this statement as we found that the concentration of sRANKL was higher in BALF of sarcoidosis patients compared with healthy subjects. sRANKL is a soluble ligand for the OPG receptor and the OPG/sRANKL system is essential for angiogenesis (Collin-Osdoby 2004). We also found an association between sRANKL and lymphocytes in BALF as both were increasing in sarcoidosis; the effect being absent in healthy subjects. These findings are, generally, in line with those of Boyce and Xing (2008) who have reported that sRANKL is expressed in activated T lymphocytes, lymph nodes, and in the lower respiratory tract and with Sandberg et al. (2006) who have found a higher expression of RANKL in T cells. Thus, it seems that T lymphocytes are an essential source of sRANKL increase. Our other finding that the level of OPG in BALF was similar in sarcoidosis patients and healthy

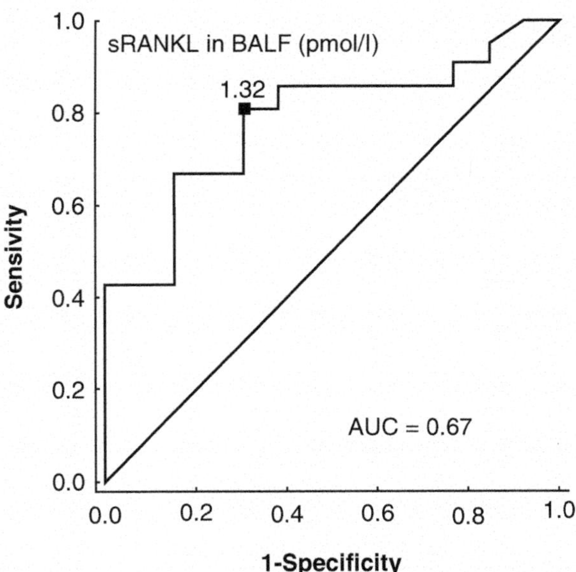

Fig. 4 Receiver operating characteristic (*ROC*) curve for the soluble receptor activator of nuclear factor-kappaB (*sRANKL*) in bronchoalveolar lavage fluid (*BALF*) for differentiation between sarcoidosis (*BBS* – Besniera-Boeck-Schaumann disease) patients and healthy subjects; the sRANKL cut-off value was 1.32 pmol/l

subjects is somehow at variance with those studies that demonstrate that OPG is a marker for cardiovascular disease risk and a predictor for cardiovascular morbidity and mortality in humans (Lieb et al. 2010). A lack of the observable increase in OPG could stem from its binding to the sRANKL.

The OPG/RANKL system is expressed in the vascular bed, including the endothelium, and is modulated by pro-inflammatory cytokines (Collin-Osdoby 2004) whose action is part of granuloma formation in BBS (Müller-Quernheim 1998). However, chronic course of pulmonary sarcoidosis leads to lung dysfunction due to fibrosis; the process involving the transforming growth factor beta (TGF-β) and IL-18 signaling pathways (Piotrowski et al. 2015; Kieszko et al. 2007). Lung dysfunction is bound to decrease lung diffusion capacity. In the present study we found that a higher level of OPG is associated with a lower DLCO, which points to a plausibly negative role for OPG in lung function.

Several studies have reported that IL-18 is closely related to the pathogenesis of pulmonary sarcoidosis by linking inflammatory immune responses and angiogenesis (Amin et al. 2010; Shigehara et al. 2001). The present findings are in line with those studies demonstrating that the concentration of IL-18 in BALF was higher in sarcoidosis patients than in healthy subjects. IL-18 in sarcoidosis patients also was positively associated with sRANKL, lending support for a connection between inflammatory responses and angiogenesis. In conclusion, the OPG/sRANKL system may be useful in clinical evaluation of sarcoidosis patients. This system raises a clinical and research interest in the still enigmatic disease which the pulmonary sarcoidosis remains.

Conflicts of Interest The authors had no conflicts of interest to declare in relation to this article.

References

American Thoracic Society (1995) Standardization of spirometry 1994 update. Am J Respir Crit Care Med 152:1107–1136

Amin MA, Rabquer BJ, Mansfield PJ, Ruth JH, Marotte H, Haas CS, Reamer EN, Koch AE (2010) Interleukin 18 induces angiogenesis *in vitro* and *in vivo* via Src and Jnk kinases. Ann Rheum Dis 69:2204–2212

Boyce BF, Xing L (2008) Functions of RANKL/RANK/OPG in bone modeling and remodeling. Arch Biochem Biophys 473:139–146

Chen ES, Moller DR (2008) Etiology of sarcoidosis. Clin Chest Med 29:365–377

Collin-Osdoby P (2004) Regulation of vascular calcification by osteoclast regulatory factors RANKL and osteoprotegerin. Circ Res 95:1046–1057

Costabel U, Hunninghake GW (1999) ATS/ERS/WASOG statement on sarcoidosis. Sarcoidosis Statement Committee. American Thoracic Society. European Respiratory Society. World Association for Sarcoidosis and Other Granulomatous Disorders. Eur Respir J 14:735–737

Cross SS, Yang Z, Brown NJ, Balasubramanian SP, Evans CA, Woodward JK, Neville-Webbe HL, Lippitt JM, Reed MW, Coleman RE, Holen I (2006) Osteoprotegerin (OPG) -a potential new role in the regulation of endothelial cell phenotype and tumour angiogenesis? Int J Cancer 118:1901–1908

Crowley LE, Herbert R, Moline JM, Wallenstein S, Shukla G, Schechter C, Skloot GS, Udasin I, Luft BJ, Harrison D, Shapiro M, Wong K, Sacks HS, Landrigan PJ, Teirstein AS (2011) 'Sarcoid like' granulomatous pulmonary disease in World Trade Center disaster responders. Am J Ind Med 54:175–184

Goswami S, Sharma-Walia N (2015) Osteoprotegerin secreted by inflammatory and invasive breast cancer cells induces aneuploidy, cell proliferation and angiogenesis. BMC Cancer 15:935

Kieszko R, Krawczyk P, Jankowska O, Chocholska S, Król A, Milanowski J (2007) The clinical significance of interleukin 18 assessment in sarcoidosis patients. Respir Med 101:722–728

Lieb W, Gona P, Larson MG, Massaro JM, Lipinska I, Keaney JF, Rong J, Corey D, Hoffmann U, Fox CS, Vasan RS, Benjamin EJ, O'Donnell CJ, Kathiresan S (2010) Biomarkers of the osteoprotegerin pathway: clinical correlates, subclinical disease, incident cardiovascular disease, and mortality. Arterioscler Thromb Vasc Biol 30:1849–1854

Liu DH, Yao YT, Cui W, Chen K (2010) The association between interleukin-18 and pulmonary sarcoidosis: a meta-analysis. Scand J Clin Lab Invest 70:428–432

Müller-Quernheim J (1998) Sarcoidosis: immunopathogenetic concepts and their clinical application. Eur Respir J 12:716–738

Piotrowski WJ, Kiszałkiewicz J, Górski P, Antczak A, Górski W, Pastuszak-Lewandoska D, Migdalska-Sęk M, Domańska-Senderowska D, Nawrot E, Czarnecka KH, Kurmanowska Z, Brzeziańska-Lasota E (2015) Immunoexpression of TGF-β/Smad and VEGF-A proteins in serum and BAL fluid of sarcoidosis patients. BMC Immunol 16:58

Reinhard H, Lajer M, Gall MA, Tarnow L, Parving HH, Rasmussen LM, Rossing P (2010) Osteoprotegerin and mortality in type 2 diabetic patients. Diabetes Care 33:2561–2566

Sandberg WJ, Yndestad A, Oie E, Smith C, Ueland T, Ovchinnikova O, Robertson AK, Muller F, Semb AG, Scholz H, Andreassen AK, Gullestad L, Damas JK, Froland SS, Hansson GK, Halvorsen B, Aukrust P (2006) Enhanced T-cell expression of RANK ligand in acute coronary syndrome: possible role in plaque destabilization. Arterioscler Thromb Vasc Biol 26:857–863

Semb AG, Ueland T, Aukrust P, Wareham NJ, Luben R, Gullestad L, Kastelein JJ, Khaw KT, Boekholdt SM (2009) Osteoprotegerin and soluble receptor activator of nuclear factor-kappaB ligand and risk for coronary events: a nested case-control approach in the prospective EPIC-Norfolk population study 1993–2003. Arterioscler Thromb Vasc Biol 29:975–980

Shigehara K, Shijubo N, Ohmichi M, Takahashi R, Kon S, Okamura H, Kurimoto M, Hiraga Y, Tatsuno T, Abe S, Sato N (2001) IL-12 and IL-18 are increased and stimulate IFN-gamma production in sarcoid lungs. J Immunol 166:642–649

Valeyre D, Prasse A, Nunes H, Uzunhan Y, Brillet PY, Müller-Quernheim J (2014) Sarcoidosis. Lancet 9923:1155–1167

Zielonka TM, Demkow U, Białas B, Filewska M, Zycinska K, Radzikowska E, Szopiński J, Skopińska-Różewska E (2007) Modulatory effect of sera from sarcoidosis patients on mononuclear cell-induced angiogenesis. J Physiol Pharmacol 58:753–766

Advs Exp. Medicine, Biology - Neuroscience and Respiration (2017) 27: 9–17
DOI 10.1007/5584_2016_55

Published online: 13 August 2016

Ambient PM2.5 Exposure and Mortality Due to Lung Cancer and Cardiopulmonary Diseases in Polish Cities

Artur J. Badyda, James Grellier, and Piotr Dąbrowiecki

Abstract

Air pollution, one of ten most important causes of premature mortality worldwide, remains a major issue also in the EU, with more than 400,000 premature deaths due to exposure to $PM_{2.5}$ reported yearly. The issue is particularly significant in Poland, where there is the highest concentration of $PM_{2.5}$ among the UE countries. This study focused on the proportion of mortality due to lung cancer and cardiopulmonary diseases attributable to $PM_{2.5}$ in eleven biggest Polish cities in the years 2006–2011. The findings demonstrate that the mean annual concentration of $PM_{2.5}$ varied from 14.3 to 52.5 $\mu g/m^3$. The average population attributable fractions varied from 0.195 to 0.413 in case of lung cancer and from 0.130 to 0.291 for cardiopulmonary diseases. Such substantial values of this ratio translate into a considerable number of deaths, which ranged between 9.6 and 22.8 cases for lung cancer and 48.6 to 136.6 cases for cardiopulmonary diseases per 100,000 inhabitants. We conclude that the impact of $PM_{2.5}$ concentration on the incidence of premature deaths is unduly high in Polish cities.

The original version of this chapter has been revised. An erratum to this chapter can be found at DOI 10.1007/978-3-319-44488-8_11

A.J. Badyda (✉)
Faculty of Environmental Engineering, Warsaw University of Technology, 20 Nowowiejska Street, 00-653 Warsaw, Poland

Central Clinical Hospital of the Ministry of National Defense, Military Institute of Medicine, 128 Szaserów Street, 04-141 Warsaw, Poland
e-mail: artur.badyda@is.pw.edu.pl

J. Grellier
Center for Research in Environmental Epidemiology, Parc de Recerca Biomèdica de Barcelona, 88 Doctor Aiguader Street, 08003 Barcelona, Spain

Department of Epidemiology and Biostatistics, Imperial College, London, UK

P. Dąbrowiecki
Central Clinical Hospital of the Ministry of National Defense, Military Institute of Medicine, 128 Szaserów Street, 04-141 Warsaw, Poland

Polish Federation of Asthma, Allergy and COPD Patients' Associations, 23/102 Łabiszyńska Street, 03-204 Warsaw, Poland

Keywords
Air pollution • Cardiopulmonary disease • City • Lung cancer • Particulate matter • Pollutant concentration • Population attributable fraction • Premature death

1 Introduction

The European Environment Agency (EEA) estimates that 9–30 % of the urban population of the EU was exposed to the concentrations of particulate matter with an aerodynamic diameter of less than 10 μm (PM_{10}) and less than 2.5 μm ($PM_{2.5}$) exceeding the upper limit values according to the 2008/50/EC Directive of the European Parliament and of the Council on ambient air quality and cleaner air for Europe (CAFE Directive) in the period of 2011–2013. That proportion has been since declining. Nonetheless, in view of the World Health Organization (WHO 2006), proportion of the urban population of the EU exposed to particulate matter exceeding the health-safe level still remains at a disquieting 61–93 % (EEA 2015).

Exposure to high levels of air pollution is associated with a wide range of acute and chronic diseases, especially of respiratory, cardiovascular, and central nervous systems (Kalkbrenner et al. 2015; WHO 2014a; Volk et al. 2013; Balmes 2009), or rheumatoid arthritis (Essouma and Noubiap 2015). Urban populations are particularly affected by adverse effects resulting from breathing highly polluted air. That creates severe social and economic problems, given that approximately 75 % of the European population live in cities (Crosette 2010). It is estimated that particulate pollutants are responsible for about 8 % of lung cancer deaths, 5 % of cardiovascular diseases, and 3 % of respiratory infections (WHO 2009). 3.7 MM premature deaths have been attributed to atmospheric pollution worldwide in both urban and non-urban populations in 2012 (WHO 2014b). Lelieveld et al. (2015) have demonstrated that air pollution, namely $PM_{2.5}$, is responsible globally for 3.3 MM deaths, with the largest impact on premature mortality in China and India, where air pollutants are emitted, in a substantial part, from the residential sources. In the EU, the number of premature deaths attributable to exposure to $PM_{2.5}$ exceeded 400,000 in 2015; a figure that was some 30,000 lower than that a year before (EEA 2015). Although the number of deaths attributable to exposure to $PM_{2.5}$ decreases in Europe as a whole, an increase in this number has been noted in countries where the problem of air pollution is particularly large, such as Poland or Bulgaria. Moreover, 16,000 premature deaths have been attributed to exposure to tropospheric ozone in the EU, the figure remaining unchanged compared with years past, and more than 70,000 deaths have been attributed to exposure to nitrogen dioxide.

The International Agency for Research on Cancer (IARC), after careful analysis of the available literature on both epidemiologic and mechanistic studies, stated in November 2013 that there is sufficient evidence on the relationship between exposure to air pollution and the incidence of lung and bladder cancers. In this context, particulate matter, a major component of air pollutants, has been classified as Group 1 carcinogen (carcinogenic to humans) (IARC 2013). Therefore, the aim of this study was to estimate the proportion of mortality from lung cancer and cardiopulmonary diseases that can be attributed to exposure to $PM_{2.5}$ in ambient air in selected Polish cities.

2 Methods

The Institutional Review Board approved this study. The consent requirement was waived because of the retrospective nature of the analysis of medical records of deceased persons. Data from the years 2006–2011 on the concentration of particulate matter in the ambient air (both $PM_{2.5}$ and PM_{10} fractions) were obtained from the General Inspectorate of Environmental Protection which is a repository of data collected by the State Environmental Monitoring agency. Aggregated annual mortality data for 11 urban agglomerations in Poland, consisting of cities with a population above 250,000 inhabitants, were acquired from the reports of the Department of Epidemiology of the Center of Oncology (lung cancer) and from the Central Statistical Office Local Data Bank (cardiopulmonary and all-cause mortality).

Exposure to air pollution in each urban agglomeration (hereafter referred to as 'city') was assessed as the mean 1-h concentration of $PM_{2.5}$, subsequently aggregated to 1-year mean values. The $PM_{2.5}$ concentration was not measured at some air monitoring stations in all relevant years. In those cases, exposure to $PM_{2.5}$ was estimated using a conversion factor determined from the results of joint $PM_{2.5}$ and PM_{10} measurements at the same station (if available) or different stations located in the same city. Depending on city, conversion factor ranged from 0.61 to 0.84.

To calculate the burden of mortality that could be attributed to $PM_{2.5}$ (population attributable fraction based on relative risk of mortality arising from exposure to $PM_{2.5}$), the exposure-response functions presented in the report on the European Perspectives on the Environmental Burden of Diseases were used (Hänninen and Knol 2011). The measures of relative risk normalized for unit exposure to $PM_{2.5}$ in case of lung cancer and cardiopulmonary diseases were taken from the results of an American study by Pope et al. (2002); those for all-cause mortality were taken from a report of the World Health Organization (WHO 2013).

3 Results

The annual mean $PM_{2.5}$ concentration in the period of 2006–2011 ranged from 14.3 to 52.5 μg/m^3 (Fig. 1), reaching the highest values in cities of southern Poland (Cracow, Katowice), where the emission of air pollutants results primarily from municipal and household sources (so-called low-stack emission) and from road transport and heavy industry. The lowest concentration was observed in cities of eastern (Bialystok, Lublin) and northern Poland (Gdansk, Szczecin), where the density of emission sources is lower and there are better climatic conditions for dispersion of air pollutants. The mean annual concentration of $PM_{2.5}$ and PM_{10} exceeded the limit level according to the CAFE Directive in 8 and 7 out of the 11 cities, respectively. In 5 of the cities, concentration of $PM_{2.5}$ and PM_{10} in 2011 increased in comparison with that in 2006.

Relative risk of mortality associated with exposure to $PM_{2.5}$, calculated for individual years and urban areas, was used to estimate the fraction of the population the deaths of who would have been prevented had the $PM_{2.5}$ concentration been reduced to the hypothetical zero level. The mean values of this population attributable fraction of all-cause mortality, and mortality from lung cancer and cardiopulmonary diseases in the years 2006–2011 demonstrate distribution and variability patterns akin to the mean concentrations of $PM_{2.5}$ shown in Fig. 1, with the highest value in southern Poland and lowest in eastern and northern Poland. Data on the population attributable fraction of mortality are shown in detail in Table 1.

The results of the population attributable fraction and the annual number of deaths from the health causes above outlined were used to estimate the number of deaths attributable to $PM_{2.5}$ in ambient air. For this purpose the number of inhabitants living in each city also was used. The estimate shows that most $PM_{2.5}$-attributable deaths occurred in Warsaw, Cracow, and Lodz. Data on $PM_{2.5}$-attributable deaths of lung cancer and cardiopulmonary diseases are shown in detail in Figs. 2 and 3.

Considering the variability of $PM_{2.5}$ level in different cities, for the sake presentation clarity, the incidence of deaths attributed to $PM_{2.5}$ was normalized for 100,000 inhabitants of each city. Table 2 presents the average population of cities of interest and Fig. 4 depicts the incidence of deaths due to all natural causes, lung cancer,

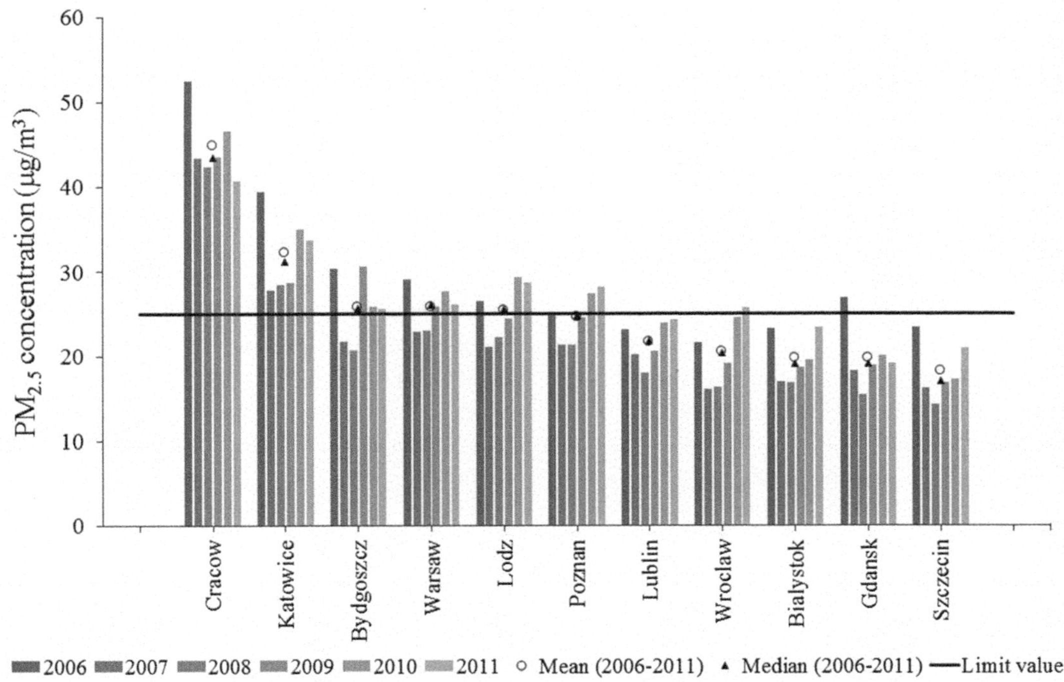

Fig. 1 Annual average concentration of $PM_{2.5}$ in Polish cities in the period of 2006–2011; limit value marked by a *horizontal line* is 25 mg/m^3

Table 1 Population attributable fractions for mortality from lung cancer, cardiopulmonary diseases, and total mortality (non-violent) in Polish cities in the period of 2006–2011

City	Lung cancer	Cardio-pulmonary diseases	All-cause (natural)
Cracow	0.413 (0.391, 0.437)	0.291 (0.278, 0.304)	0.242 (0.233, 0.251)
Katowice	0.318 (0.295, 0.342)	0.218 (0.207, 0.230)	0.180 (0.172, 0.189)
Bydgoszcz	0.265 (0.246, 0.284)	0.179 (0.170, 0.189)	0.147 (0.141, 0.154)
Warsaw	0.264 (0.254, 0.276)	0.179 (0.174, 0.185)	0.147 (0.144, 0.151)
Lodz	0.261 (0.246, 0.277)	0.177 (0.170, 0.184)	0.145 (0.140, 0.150)
Poznan	0.254 (0.242, 0.268)	0.172 (0.166, 0.178)	0.141 (0.137, 0.146)
Lublin	0.228 (0.218, 0.239)	0.153 (0.149, 0.158)	0.126 (0.122, 0.129)
Wroclaw	0.217 (0.201, 0.234)	0.146 (0.138, 0.154)	0.119 (0.114, 0.125)
Bialystok	0.210 (0.199, 0.222)	0.141 (0.136, 0.146)	0.115 (0.112, 0.119)
Gdansk	0.210 (0.196, 0.225)	0.141 (0.134, 0.148)	0.115 (0.110, 0.120)
Szczecin	0.195 (0.183, 0.207)	0.130 (0.124, 0.136)	0.106 (0.102, 0.110)

Data are mean values (95 % confidence intervals)

and cardiopulmonary diseases attributable to $PM_{2.5}$.

The incidence of deaths per 100,000 inhabitants attributable to $PM_{2.5}$ exposure, particularly concerning lung cancer and cardiopulmonary diseases, was the highest where the concentration of this pollutant was the highest as well (Cracow and Katowice). A comparably high incidence of deaths was recorded in Lodz; where $PM_{2.5}$ concentration was noticeably lower. Thus, other factors, e.g., age, could play a role in increasing the number of $PM_{2.5}$-attributable deaths in case of Lodz. Moreover, a high mortality rate in this city could reflect a high residual concentration of air pollutants due to a spate of light industry Lodz had been renowned for over the years past. The lowest mortality rate was noted in the eastern city of Bialystok, and a

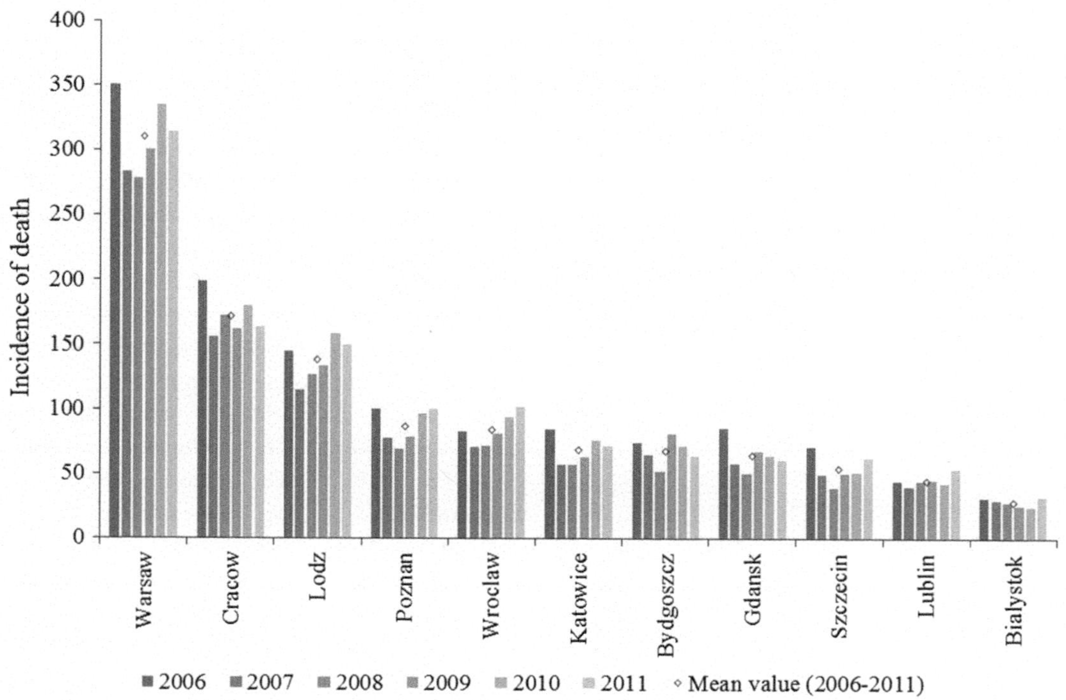

Fig. 2 Lung cancer mortality attributable to ambient $PM_{2.5}$ in Polish cities in the period of 2006–2011

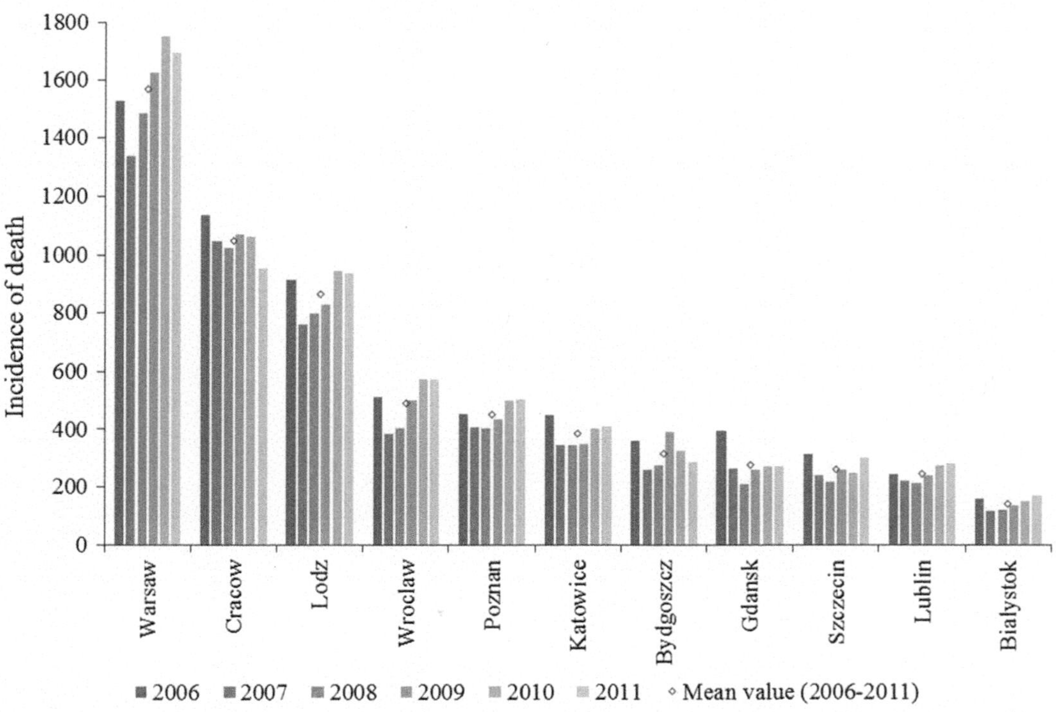

Fig. 3 Cardiopulmonary mortality attributable to ambient $PM_{2.5}$ in Polish cities in the period of 2006–2011

slightly greater level was in the northern Baltic port cities of Szczecin and Gdansk.

4 Discussion

This study presents estimates of a total number of deaths from all natural causes, and separately caused by lung cancer and cardiopulmonary diseases, that can be attributed to exposure to $PM_{2.5}$. The results indicate that $PM_{2.5}$ has a noticeable impact on the number of deaths. It was clearly observed that in cities characterized by the highest levels of particulate matter (Cracow, Katowice) shares of deaths attributed to $PM_{2.5}$ against the background of the total number of deaths are the highest.

While the results of this study provide very useful information for policy makers regarding the potential impact of air pollution on public health in Poland, a relatively simple approach used to calculating the attributable cases has some limitations. The same exposure to concentrations of $PM_{2.5}$, to start with, was assigned to all inhabitants of a given city, thereby ignoring the intra-city variation in the true exposure that existed. This simplification was made due to a limited number of air quality monitoring stations in each of the cities of interest (in some cities monitoring results were available for only one station) and due to the lack of possibility of

Table 2 Average number of inhabitants in individual cities in the period of 2006–2011

City	Population
Warsaw	1,706,932
Cracow	756,559
Lodz	743,112
Wroclaw	632,299
Poznan	557,758
Gdansk	457,596
Szczecin	408,328
Bydgoszcz	361,455
Lublin	350,540
Katowice	310,933
Bialystok	294,377

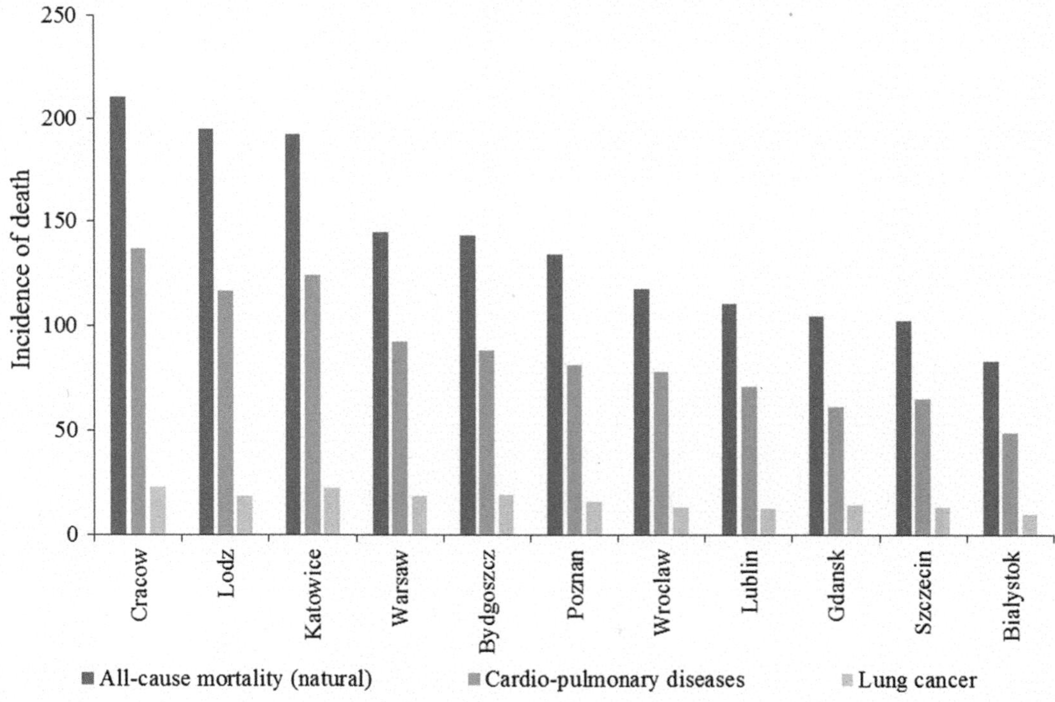

Fig. 4 Average incidence of all-cause, lung cancer, and cardiopulmonary mortality per 100,000 inhabitants attributable to ambient $PM_{2.5}$ in Polish cities in the period of 2006–2011

referring specific and differentiated concentrations of $PM_{2.5}$ to certain fraction of the population living in these cities. In carrying out the analysis, this lack in sufficient density of sampling stations was a finding of some considerable importance in terms of environmental and public health policy as only with adequate network of air pollution monitors can a more complete picture of the impact of air pollution on the health of Polish citizens be understood. In case of achieving the ability to expand the air pollution monitoring system in cities, it would be possible to use the tools for modeling distributions of $PM_{2.5}$ in the agglomeration, which can help in getting more reliable results.

Due to the use of aggregated mortality data, there was no way of estimating the impact of effect modification due to tobacco use (see Künzli et al. 2005) or other potentially important covariates, e.g., socioeconomic status. Furthermore, relative risks normalized for a unit exposure to $PM_{2.5}$ were used, which are derived from studies conducted mostly in North American populations (Krewski et al. 2009). It is likely that these relative risks should not be transferred directly for use in Polish populations without recognizing the potential for introducing bias due to differences in distributions of (unmeasured) covariates such as tobacco use, socioeconomic status, ethnicity, and others. However, due to the lack of epidemiological studies in this field in Poland it is currently not possible to verify the magnitude and direction of such a bias.

Taking into account the generally high levels of particulate matter pollution in Polish cities, which is one of the highest among all EU countries, and exceedances not only of the restrictive WHO guidelines regarding the recommended concentrations of $PM_{2.5}$ in the ambient air, but also much more liberal limits under the EU and national law, it should be noted that the contribution of the risk factor related to the impact of air pollution on mortality is relatively high.

Although there are studies indicating the lack of significant relationships between exposure to particulate matter and cardiovascular (Wang et al. 2014) or respiratory (Dimakopoulou et al. 2014) diseases mortality, most of the research in this area indicates an increased mortality risk for those exposed to long-term and short-term impact of air pollution (including PM_{10} and $PM_{2.5}$). In many studies on similar issues, which were carried out in other countries, significantly lower concentrations of $PM_{2.5}$ in ambient air were found. The assessment of the influence of air pollutants on mortality in 22 cohort studies carried out in Europe shows that $PM_{2.5}$ concentrations ranged from 6.6 to 31.0 $\mu g/m^3$ (Beelen et al. 2013). When compared with the cities of southern Poland, as demonstrated by the results presented in this paper, exposure of subjects in those cohorts was significantly lower. Nonetheless, there was a 1.07 increase in relative risk of mortality attributed to each increase of $PM_{2.5}$ concentration by each 5 $\mu g/m^3$, while pointing out that this pollutant was most closely associated with mortality in relation to all other types of air pollutants investigated (especially PM_{10}, NO_2, and NO_x). An increasing risk of mortality for lung cancer and stroke, but not for ischemic heart disease and respiratory diseases, has also been observed. Analyses regarding the same cohorts have also revealed a significant association between increasing concentration of $PM_{2.5}$ and relative risk of mortality from cerebrovascular diseases, growing by 1.21 with increased $PM_{2.5}$ concentration by each 5 $\mu g/m^3$ (Beelen et al. 2014). A Dutch study on long-term exposure on traffic-related air pollutants and its association with mortality indicated that each 10 $\mu g/m^3$ increase in $PM_{2.5}$ concentration is associated with higher relative risk (RR) of mortality due to natural causes (RR = 1.06), cardiovascular diseases (RR = 1.04), and respiratory diseases (RR = 1.07) (Beelen et al. 2008). In that study even slightly stronger associations with NO_2 (each 30 $\mu g/m^3$ concentration increase) were also ascertained. In a Brazilian study, a 3.3 %, 3.8 %, and 6.0 % increases in daily mortality for all causes, cardiovascular, and respiratory diseases, respectively, for an increase of PM_{10} concentration from the 10th to 90th percentile have been reported (Gouveia and Fletcher 2000). In turn, research conducted in the US shows an increase in total mortality risk of 1.06 with increasing $PM_{2.5}$ concentration by 10 μg/m^3 (Pope et al. 2002) and in a Canadian cohort increased risk of natural deaths (RR = 1.15) and ischemic heart disease (RR = 1.31) have been assigned to the same growth of $PM_{2.5}$ concentration (Crouse et al. 2012). A 6 % increase in total mortality risk and 11 % in cardiopulmonary mortality (ischemic heart disease in particular) have been revealed in long-term exposure to $PM_{2.5}$ pollution, covering many areas of the world, including Asian countries (Hoek et al. 2013). In a Harvard Six Cities study (Laden et al. 2006) and a US study of 36 cities (Miller et al. 2007), significantly higher risks (28 % and 76 %, respectively) of death due to cardiopulmonary diseases have been found associated with increased $PM_{2.5}$ concentration by 10 μg/m^3. However, a British cohort study demonstrates no appreciable change of risk of death from cardiopulmonary diseases dependent on changes in air quality (Carey et al. 2013). Shah et al. (2013) show a slight increase in relative risk of hospitalization or death due to circulatory failure associated with growing concentration of $PM_{2.5}$ (about 1.02 per 10 μg/m^3 of $PM_{2.5}$). Considerably higher relative risks of morality are observed in studies carried out in China, where the problem of air pollution is one of the largest in the world. A study by Dong et al. (2012) demonstrates that relative risk of death due to respiratory diseases is 1.67 per 10 μg/m^3 increase in PM_{10}. Other Chinese research has shown growing mortality outcomes associated with increased concentration of PM_{10}. A 10 μg/m^3 increase of PM_{10} concentration causes an increase of total, cardiovascular, and respiratory mortality by 25 %, 27 %, and 27 %, respectively. The effects are similar in cool and warm seasons except for respiratory mortality which is considerably higher in winter (Kan et al. 2008).

The present study focused only on the 11 largest cities in Poland, as these are the ones currently monitored for air pollution on a routine basis. Given the presence of heavy industry and the use of low quality coal and other poor quality fuels for domestic heating in some of the many smaller cities in Poland, in particular those in the densely populated southwestern part of the country, it seems likely that the current study results present a considerable underestimate of the public health impacts of air pollution in the country as whole.

The results presented in this article represent a preliminary study on the assessment of the health influence, in terms of mortality, due to fine particulate air pollution in Poland. The potential impact of fine particulate air pollution on public health in Poland is much higher relative to the other member states of the EU. Deleterious health effects are associated with considerable costs in both social and economic spheres. This study serves to indicate both the very limited information available for the accurate assessment of environmental quality of the majority of Polish cities and the lack of a developed infrastructure for assessing risks presented to the Polish population from environmental contaminants. The development of an improved integrated network of air pollution monitoring and funding epidemiological research toward a better understanding of the environmental detriments to health in Poland would also serve to improve the quality of subsequent health assessment studies on air pollution, something which would ultimately help policy makers to reduce emissions, improve air quality, and benefit public health.

Conflicts of Interest The authors declare no conflicts of interest in relation to this article.

References

Balmes JR (2009) Can traffic-related air pollution cause asthma? Thorax 64:646

Beelen R, Hoek G, van den Brandt PA, Goldbohm RA, Fischer P, Schouten LJ, Jerrett M, Hughes E, Armstrong B, Brunekreef B (2008) Long-term effects of traffic-related air pollution on mortality in a Dutch cohort (NLCS-AIR study). Environ Health Perspect 116(2):196–202

Beelen R, Raaschou-Nielsen O, Stafoggia M et al (2013) Effects of long-term exposure to air pollution on natural-cause mortality: an analysis of 22 European cohorts within the multicentre ESCAPE project. Lancet 383(9919):785–795

Beelen R, Stafoggia M, Raaschou-Nielsen O et al (2014) Long-term exposure to air pollution and cardiovascular mortality. An analysis of 22 European cohorts. Epidemiology 25(3):368–378

Carey IM, Atkinson RW, Kent AJ, van Staa T, Cook DG, Anderson HR (2013) Mortality associations with long-term exposure to outdoor air pollution in a national English cohort. Am J Respir Crit Care Med 187:1226–1233

Crosette B (2010) State of the world population 2010. New York, United Nations Population Fund

Crouse DL, Peters PA, van Donkelaar A, Goldberg MS, Villeneuve PJ, Brion O, Khan S, Atari DO, Jerrett M, Pope CA, Brauer M, Brook JR, Martin RV, Stieb D, Burnett RT (2012) Risk of nonaccidental and cardiovascular mortality in relation to long-term exposure to low concentrations of fine particulate matter: a Canadian national-level cohort study. Environ Health Perspect 120(5):708–714

Dimakopoulou K, Samoli E, Beelen R et al (2014) Air pollution and nonmalignant respiratory mortality in 16 cohorts within the ESCAPE project. Am J Respir Crit Care Med 189(6):684–696

Dong GH, Zhang P, Sun B, Zhang L, Chen X, Ma N, Yu F, Guo H, Huang H, Lee YL, Tang N, Chen J (2012) Long-term exposure to ambient air pollution and respiratory disease mortality in Shenyang, China: a 12-year population-based retrospective cohort study. Respiration 84(5):360–368

EEA (2015) European Environment Agency. Air quality in Europe – 2015 report. Publications Office of the European Union, ISSN 1977–8449. Luxembourg

Essouma M, Noubiap JJN (2015) Is air pollution a risk factor for rheumatoid arthritis? J Inflamm (Lond) 12:48. doi:10.1186/s12950-015-0092-1

Gouveia N, Fletcher T (2000) Time series analysis of air pollution and mortality: effects by cause, age and socioeconomic status. J Epidemiol Community Health 54(10):750–755

Hänninen O, Knol A (2011) European perspectives on environmental burden of diseases; estimates for nine stressors in six European countries. National Institute for Health and Welfare, Report 1/2011, ISBN: 978-952-245-413-3. Helsinki

Hoek G, Krishnan MK, Beelen R, Peters A, Ostro B, Brunekreef B, Kaufman JD (2013) Long-term air pollution exposure and cardio-respiratory mortality. Environ Health 12:43

IARC (2013) International Agency for Research on Cancer. Outdoor air pollution a leading environmental cause of cancer deaths, Press release N°221 from: http://www.iarc.fr/en/media-centre/pr/2013/pdfs/pr221_E.pdf. Accessed on Jan 2016

Kalkbrenner AE, Windham GC, Serre ML, Akita Y, Wang X, Hoffman K, Thayer BP, Daniels JL (2015) Particulate matter exposure, prenatal and postnatal windows of susceptibility, and autism spectrum disorders. Epidemiology 26(1):30–42

Kan H, London SJ, Chen G, Zhang Y, Song G, Zhao N, Jiang L, Chen B (2008) Season, sex, age, and education as modifiers of the effects of outdoor air pollution on daily mortality in Shanghai, China: the public health and air pollution in Asia (PAPA) Study. Environ Health Perspect 116(9):1183–1188

Krewski D, Jerrett M, Burnett RT et al. (2009) Extended follow-up and spatial analysis of the American Cancer Society study linking particulate air pollution and mortality. Health Research Institute, Research Report 140/2009. Boston

Künzli N, Jerrett M, Mack WJ, Beckerman B, LaBree L, Gilliland F, Thomas D, Peters J, Hodis HN (2005) Ambient air pollution and atherosclerosis in Los Angeles. Environ Health Perspect 113(2):201–206

Laden F, Schwartz J, Speizer FE, Dockery DW (2006) Reduction in fine particulate air pollution and mortality: extended follow-up of the Harvard Six Cities study. Am J Respir Crit Care Med 173:667–672

Lelieveld J, Evans JS, Fnais M, Giannadaki D, Pozzer A (2015) The contribution of outdoor air pollution sources to premature mortality on a global scale. Nature 525(7569):367–371

Miller KA, Siscovick DS, Sheppard L, Shepherd K, Sullivan JH, Anderson GL, Kaufman JD (2007) Long-term exposure to air pollution and incidence of cardiovascular events in women. N Engl J Med 356:447–458

Pope CA 3rd, Burnett RT, Thun MJ, Calle EE, Krewski D, Thurston GD (2002) Lung cancer, cardiopulmonary mortality, and long-term exposure to fine particulate air pollution. JAMA 287(9):1132–1141

Shah AS, Langrish JP, Nair H, McAllister DA, Hunter AL, Donaldson K, Newby DE, Mills NL (2013) Global association of air pollution and heart failure: a systematic review and meta-analysis. Lancet 382 (9897):1039–1048

Volk HE, Lurmann F, Penfold B, Hertz-Picciotto I, McConnell R (2013) Traffic-related air pollution, particulate matter, and autism. JAMA Psychiatry 70 (1):71–77

Wang M, Beelen R, Stafoggia M et al (2014) Long-term exposure to elemental constituents of particulate matter and cardiovascular mortality in 19 European cohorts: results from the ESCAPE and TRANSPHORM projects. Environ Int 66:97–106

WHO (2006) Air quality guidelines for particulate matter, ozone, nitrogen dioxide and sulfur dioxide. Global update 2005. Geneva

WHO (2009) Global health risk. Mortality and burden of disease attributable to selected major risk. WHO Library Cataloguing-in-Publication Data, ISBN: 978-92-4-156387-1

WHO (2013) Health risks of air pollution in Europe – HRAPIE project. Copenhagen

WHO (2014a) Burden of disease from ambient air pollution for 2012. Geneva

WHO (2014b) The top 10 causes of death. Fact sheet Nr 310

Advs Exp. Medicine, Biology - Neuroscience and Respiration (2017) 27: 19–25
DOI 10.1007/5584_2016_45

Published online: 12 July 2016

Non-Tuberculous Mycobacteria: Classification, Diagnostics, and Therapy

I. Porvaznik, I. Solovič, and J. Mokrý

Abstract

Non-tuberculous mycobacteria (NTM) are species other than those belonging to the *Mycobacterium tuberculosis* complex and do not cause leprosy. NTM are generally free-living organisms that are ubiquitous in the environment. There have been more than 140 NTM species identified to-date. They can cause a wide range of infections, with pulmonary infections being the most frequent (65–90 %). There is growing evidence that the incidence of NTM lung diseases and associated hospitalizations are on the rise, mainly in regions with a low prevalence of tuberculosis. A crucial clinical problem remains the evaluation of NTM significance in relation to the disease, especially in regard to the colonization of the respiratory tract in patients with residual lesions after tuberculosis or bronchiectasis. Clinical and radiographic pictures of mycobacteriosis, as well as therapy, have often similarities to those of tuberculosis. The treatment regimen should be individualized. In addition to antituberculotics, antibiotics are used more frequently. The most common mycobacteria causing lung disease in Slovakia are *Mycobacterium avium* and *Mycobacterium abscessus*.

Keywords

Antituberculotics • Diagnosis • Infection • Lung • Mycobacteriosis • Non-tuberculous mycobacteria • Respiratory tract

The original version of this chapter has been revised. An erratum to this chapter can be found at DOI 10.1007/978-3-319-44488-8_11

I. Porvaznik (✉)
Department of Pulmonary Diseases and Thoracic Surgery, National Institute for Tuberculosis, 1 Vysne Hagy, 05984 Vysoké Tatry, Slovakia

Division of Respirology and Department of Pharmacology, Biomedical Center Martin, Jessenius School of Medicine in Martin, Comenius University, Bratislava, Martin, Slovakia
e-mail: porvaznik@hagy.sk

I. Solovič
Department of Pulmonary Diseases and Thoracic Surgery, National Institute for Tuberculosis, 1 Vysne Hagy, 05984 Vysoké Tatry, Slovakia

J. Mokrý
Division of Respirology and Department of Pharmacology, Biomedical Center Martin, Jessenius School of Medicine in Martin, Comenius University, Bratislava, Martin, Slovakia

1 Introduction and Taxonomy

Non-tuberculous mycobacteria (NTM) are species other than those belonging to the *Mycobacterium tuberculosis* complex and do not cause leprosy. NTM are generally free-living organisms that are ubiquitous in the environment. There have been more than 140 NTM species identified to-date. They can cause a wide range of mycobacterial infections, with pulmonary infections being the most frequent (65–90 %) (van Ingen et al. 2012d).

Mycobacteria are aerobic, non-motile organisms that appear positive with acid-fast alcohol staining. They have a lipid rich, hydrophobic cell wall, substantially thicker than that of the most other bacteria. The thickness and composition of the cell wall renders mycobacteria impermeable to hydrophilic nutrients and resistant to heavy metals, disinfectants, and antibiotics (Jarlier and Nikaido 1994). Four groups of human pathogens are recognized in the *Mycobacterium* genus: (1) *Mycobacterium tuberculosis* complex, (2) *Mycobacterium leprae*, (3) slowly growing NTM, and (4) rapidly growing mycobacteria.

Historically, NTM have been classified by according to their growth rate and pigment formation (Runyon 1959) (Fig. 1). Types I, II, and III, so-called slow growers take 7 days or more to grow, and have been classified by their coloration. If the pigment is produced only on exposure to light, they are photochromogens (type I); if it is produced in the dark, they are scotochromogens (type II); if the bacteria are not strongly pigmented, they are non-photochromogens (type III). Rapid growers (type IV) grow in less than 7 days, which however, is still more slowly than most other bacteria do.

2 Epidemiology

Infections caused by NTM differ on several points from the classical epidemiology of tuberculosis (TB):

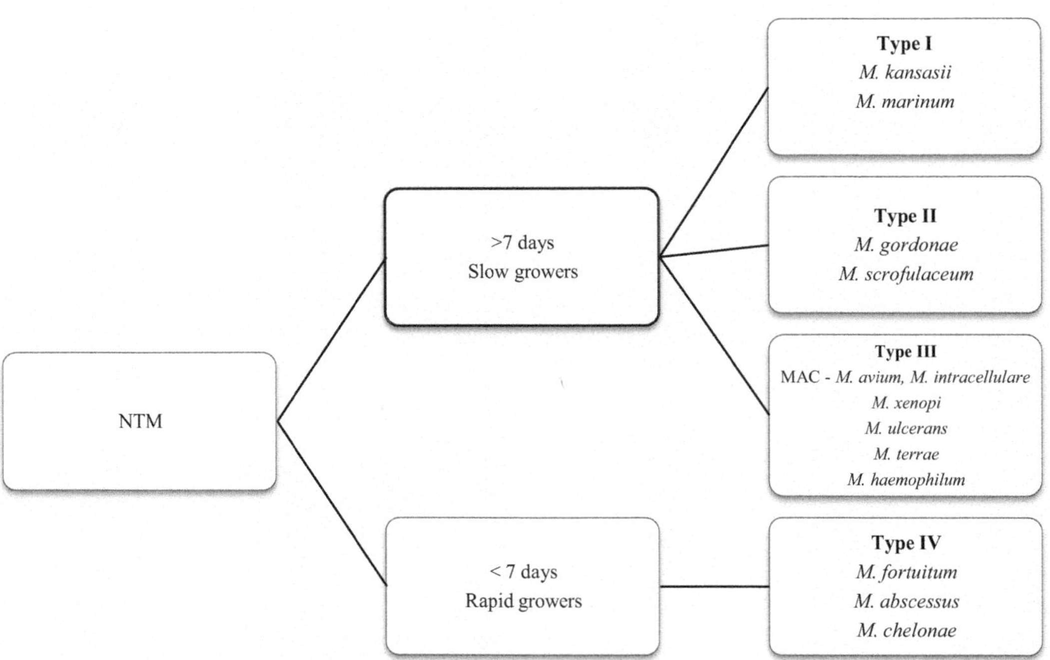

Fig. 1 Classification of non-tuberculous mycobacteria (NTM) (Adapted from Runyon 1959)

1. classical TB is widespread worldwide, NTM infections are characterized by a distinct endemic occurrence;
2. source of infection for humans in classical TB is a sick person or an infected animal, while NTM reservoirs are present in the environment;
3. route of transmission of TB is direct from the source of infection, rarely indirect from contaminated food, and the transfer mechanism is known. Transmission routes of most NTM diseases represent a problematic issue and the mechanism has not yet been adequately explained yet.

There is a growing body of evidence that the incidence of NTM lung diseases and associated hospitalizations are on the rise, mainly in the regions with a low prevalence of TB (Winthrop et al. 2010). Iseman and Marras (2008) have mentioned that new cases of NTM lung disease reach or even exceed those of pulmonary TB in a number of industrialized countries. The underlying factors of this changing epidemiology encompass an increase in the prevalence of susceptible hosts. The factors include patients requiring systemic therapy due to a severe disease, e.g., HIV infection, hematological malignancy, inheritable disorders of immunity, immunosuppressive drug use including TNF-α therapy (Winthrop et al. 2009), systemic or inhaled corticosteroid therapy (Andrejcak et al. 2013), pre-existing pulmonary disease such as cystic fibrosis, or chronic obstructive pulmonary disease (van Ingen et al. 2012c).

3 Diagnosis of NTM Pulmonary Disease

The American Thoracic Society (ATS) and the Infectious Diseases Society of America (IDSA) have recently issued a statement containing a set of criteria to differentiate an accidental non-morbid NTM isolation from a real pulmonary disease (Griffith et al. 2007):

Clinical Criteria

1. pulmonary symptoms;
2. nodular or cavitary opacities on chest radiograph, or a HRCT scan showing a multifocal bronchiectasis with multiple small nodules;
3. appropriate exclusion of other diagnoses.

All three clinical criteria are to be fulfilled to confirm the diagnosis.

Microbiological Criteria

1. positive cultivation from at least two separate expectorated sputum samples;
2. positive cultivation from at least one bronchial wash or lavage;
3. transbronchial or other lung biopsy with mycobacterial histopathological features and a positive cultivation for NTM, or biopsy showing mycobacterial histopathological features and a sputum or bronchial washing testing positive for NTM.

Only one of the microbiological criteria is required to confirm the diagnosis.

NTM should be identified at the species level. In the National Reference Laboratory for mycobacteria in Vysne Hagy in Slovakia we have been recently switching from phenotypic and biochemical analyses to molecular-genetic methods. Two kinds of tests are in use at present: (1) Speed-oligo® Mycobacteria (Vircell, Spain) which enables the identification of *Mycobacterium tuberculosis* complex and other 13 most frequent NTM and (2) Genotype Mycobacterium CM/AS (Hain Lifescience, Germany) which enables the identification of 14 common NTM and other 16 additional species (Porvaznik et al. 2015; Porvaznik et al. 2014).

4 Drug-Susceptibility Testing (DST)

The role of DST in the choice of suitable drugs for antimicrobial treatment of NTM disease (caused mainly by slow growers) remains a subject of discussions (van Ingen et al. 2012b). There are important discrepancies between minimum inhibitory concentrations (MICs) measured in *in vitro* conditions and the activity of the respective drug observed *in vivo* (van Ingen et al. 2012a). For the *Mycobacterium avium* complex, the susceptibility testing of macrolides only (i.e., clarithromycin) is currently recommended, since this testing has been clinically validated (CLSI 2011; Wallace et al. 1996). For *Mycobacterium kansasii*, the initial testing should include only rifampicin. Rifampicin resistant isolates have been observed in patients who fail to respond to treatment with rifampicin-based regimen (Wallace et al. 1994).

For rapid growers, relationship between MICs and clinical outcome has been studied for several drugs such as tobramycin, co-trimoxazole, cefoxitin, and doxycycline. However, these drugs have been tested mostly in extrapulmonary diseases and the key drugs, including amikacin and macrolides, have not been included (Wallace et al. 1985). MIC of any drug other than that mentioned above should be interpreted with caution, and an expert consultation before applying non-standard therapy is recommended.

5 Treatment of Non-Tuberculous Mycobacterial (NTM) Pulmonary Diseases

The clinical and radiographic picture of NTM as well as treatment are often akin to that of TB. However, antibiotics are used more frequently than antituberculotics. The main cause of chemotherapy failure is undoubtedly the primary resistance of NTM to most classical antituberculotics. Furthermore, time between the identification of a pathogen and commencement of therapy is rather long in most cases. Except for the pulmonary forms of the disease caused by *M. kansasii*, the long-term treatment with antituberculotics of lung lesions is ineffective.

In contrast to TB, diagnosis of NTM lung disease does not necessarily require specific treatment. The final decision concerning pharmacotherapy requires an individual approach depending on a specific NTM species, patient acceptance, tolerance and compliance, and treatment goals, for instance, a reduction of symptoms or sputum conversion. Treatment options include an observation of the disease course with the best possible pulmonary care, a course of antibiotics given in a constant or intermittent dosage regimen, and sometimes intravenous therapy for several months or surgical treatment (Griffith et al. 2007).

There are several issues typical for NTM which influence the effectiveness of antibiotic therapy. The *in vitro* susceptibility testing often does not provide a good guidance for an effective *in vivo* response to antibiotics. One of the most important therapeutic goals is to avoid the emergence of macrolide resistant *Mycobacterium avium* complex infections (Griffith et al. 2007) or *M. abscessus* (Koh et al. 2011) strains during therapy. The primary goal of treatment is a 12-month period of sputum culture negativity while on therapy. The recommended treatment regimens for selected NTM respiratory pathogens are listed in Table 1. However, these multi-drug regimens may lead to significant pharmacokinetic interactions. In particular, rifampicin reduces the serum levels of macrolides and moxifloxacin in patients with

Table 1 Recommended treatment regimens for NTM respiratory pathogens

Mycobacterial species	Drugs
M. avium complex	macrolide, rifamycins, ethambutol
M. kansasii	rifampicin, ethambutol, izoniazid
M. xenopi	macrolide, rifamycins, ethambutol
M. chelonae	aminoglycosides, macrolide
M. abscessus	amikacin, cefoxitin, imipenem, tigecycline, linezolide, macrolide
M. fortuitum	amikacin, cefoxitin, sulfonamides

NTM pulmonary disease. Clinical implications of those reductions remain unknown, but they may have to do with a poor outcome of antibiotic therapy. Treatment outcomes differ depending on the bacterial species. The best therapeutic results have been observed in *M. kansasii* and *M. malmoense* infections. Worse results concern *Mycobacterium avium* complex, and very poor outcomes have been reported in patient with pulmonary NTM disease caused by *M. xenopi*, *M. simiae,* and, particularly. *M. abscessus* subsp. *abscessus* (van Ingen et al. 2012c).

6 Actual Situation in Slovakia

The number of isolated non-tuberculous mycobacteria strains has significantly increased as registered by the National Reference Laboratory for Mycobacteria during last years (Table 2). The most frequently isolated strains include *M. gordonae,* whose cultures grow slowly and are smooth and yellow pigmented. As they are frequent isolates in the tap water, *M. gordonae* is often referred to as a tap water bacillus (Lalande et al. 2001) and should rather be interpreted as a contamination.

In 2014, rapidly growing mycobacteria have been found in clinical samples from 23 patients; there were eight cases of *M. fortuitum*, six of *M. chelonae*, another six of *M. mucogenicum*, and three of *M. abscessus* (Table 2). Based on clinical status of patients and their lung X-ray findings, only six cases were diagnosed as mycobacteriosis (Table 3). The most serious clinical issue of 2014 was represented by three cases of pulmonary mycobacteriosis caused by *M. abscessus*. One of the patients died, the other

Table 2 Non-tuberculous mycobacteria strains isolated in Slovakia registered by the National Reference Laboratory in consecutive years

Mycobacterial species	2010	2011	2012	2013	2014	2015[a]
M. gordonae	25	25	28	34	54	48
M. xenopi	6	8	4	13	3	3
M. avium	4	4	5	8	12	11
M. intracellulare	0	1	0	2	2	4
M. scrofulaceum	0	0	0	0	1	1
M. chelonae	4	3	5	4	6	7
M. abscesus	0	3	1	1	3	2
M. fortuitum	2	2	6	7	8	8
M. mucogenicum	0	0	1	6	6	3
M. kansasii	3	0	2	0	2	1
M. lentiflavum	1	0	1	0	1	3

[a]Situation by the end of November 2015

Table 3 Diagnosed lung mycobacterioses in Slovakia in consecutive years

Mycobacterial species	2010	2011	2012	2013	2014	2015[a]
M. avium	3	3	6	7	8	6
M. intracellulare	0	0	0	2	2	3
M. xenopi	3	3	4	2	2	2
M. abscesus	0	2	1	2	3	2
M. chelonae	1	1	1	1	0	0
M. srofulaceum	0	0	0	0	1	1
M. kansasii	1	0	1	1	2	1
All	8	9	13	15	18	14

[a]Situation by the end of November 2015

two have remained culture positive, despite the antibiotic therapy, suggesting that *M. abscessus* is a new scary bacterial nightmare.

Similar therapeutic results are observed with lung mycobacterioses caused by *Mycobacterium avium* complex. In these patients, radiological regression takes place while on therapy, associated with a decrease in cultivation positivity. However, reinfection occurs after a time, suggesting an insufficient antibiotic control. Wallace et al. (2014) have argued that reinfection, and not antibiotic resistance, is the major cause treatment failure. These authors achieved preliminary cure, i.e., sputum conversion without true microbiologic relapse, in 84 % of patients. Nonetheless, microbiologic recurrence occurred in 48 % of these patients after therapy completion, due to reinfection isolates (75 %) and true relapse isolates (25 %).

7 Conclusions

Non-tuberculous mycobacteria represent a relatively rare clinical and laboratory finding, but the frequency of lung infections caused by these bacteria has been steadily rising in Slovakia during in recent years. A crucial task for clinicians is to rationally evaluate the clinical significance of detected mycobacterial agents in relation to the disease symptomatology, especially in assessing the colonization of the respiratory tract in patients with residual lesions after tuberculosis or bronchiectasis.

From the microbiologic standpoint, heterogeneity of microbes requires sophisticated and rapid laboratory techniques. Since current pharmacological therapy of mycobacterioses is problematic and it often fails to reach the long-term eradication of pathogens, it is necessary to ferret out new drugs or treatment and dosage regimens for successful therapy of these infections, particularly serious in immunocompromised patients.

Conflicts of Interest The authors declare no conflicts of interest in relation to this article.

References

Andrejcak C, Nielsen RB, Thomsen V, Duhaut P, Sørensen HT, Thomsen RW (2013) Chronic respiratory disease, inhaled corticosteroids and risk of nontuberculous mycobacteriosis. Thorax 68:256–262

CLSI (2011) Susceptibility testing of *Mycobacteria, Nocardiae* and other aerobic actinomycetes; approved standard, 2nd edn. Clinical and Laboratory Standard Institute, Wayne

Griffith DE, Aksamit T, Brown-Elliot BA, Catabzaro A, Daley C, Gordin F, Holland SM, Horsburgh R, Huitt G, Iademarco MF, Iseman M, Olivier K, Ruoss S, Von Reyn CF, Wallce RJ Jr, Winthrop K, on behalf of the ATS Mycobacterial Diseases Subcomittee (2007) An official ATS/IDSA statement: diagnosis, tratment and prevention of nontuberculous mycobacterial diseases. Am J Respir Crit Care 175:367–416

Iseman MD, Marras TK (2008) The importance of nontuberculous mycobacterial disease. Am J Respir Crit Care Med 178:999–1000

Jarlier V, Nikaido H (1994) Mycobacterial cell wall: structure and role in natural resistance to antibiotics. FEMS Microbiol Lett 123:11–18

Koh WJ, Jeon K, Lee NY, Kim BJ, Kook YH, Lee SH, Park YK, Kim CK, Shin SJ, Huitt GA, Daley CL, Kwon OJ (2011) Clinical significance of differentiation of *Mycobacterium massiliense* from *Mycobacterium abscessus*. Am J Respir Crit Care Med 183:405–410

Lalande V, Barbut F, Varnerot A, Febvre M, Nesa D, Wadel S, Vincent V, Petit JC (2001) Pseudo-outbreak of *Mycobacterium gordonae* associated with water from refrigerated fountains. J Hosp Infect 48:76–79

Porvaznik I, Mokry J, Solovic I (2014) Drug resistance to anti-tuberculotics in children – three years status in Slovakia. Acta Med Martiniana 13:18–22

Porvaznik I, Mokry J, Solovic I (2015) Classical against molecular-genetic methods for susceptibility testing of antituberculotics. Adv Exp Med Biol 4:15–22

Runyon EH (1959) Anonymous mycobacteria in pulmonary disease. Med Clin North Am 43:273–290

van Ingen J, Boeree MJ, van Soolingen D, Iseman MD, Heifets LB, Daley CL (2012a) Are phylogenetic position, virulence, drug susceptibility and *in vivo* response to treatment in mycobacteria interrelated? Infect Genet Evol 12:832–837

van Ingen J, Boeree MJ, van Soolingen D, Mouton JW (2012b) Resistance mechanisms and drug susceptibility testing of nontuberculous mycobacteria. Drug Resist Updat 15:149–161

van Ingen J, Egelund EF, Levin A, Totten SE, Boeree MJ, Mouton JW, Aarnoutse RE, Heifets LB, Peloquin CA, Daley CL (2012c) The pharmacokinetics and pharmacodynamics of pulmonary *Mycobacterium avium* complex disease treatment. Am J Respir Crit Care Med 186:559–565 van Ingen J, Griffith DE, Aksamit TR, Wagner D (2012d) Pulmonary diseases caused by non-tuberculous mycobacteria. Eur Respir Monogr 58:25–37

Wallace RJ Jr, Swenson JM, Silcox VA, Bullen MG (1985) Treatment of nonpulmonary infections due to *Mycobacterium fortuitum* and *Mycobacterium chelonei* on the basis of in vitro susceptibilities. J Infect Dis 152:500–514

Wallace RJ Jr, Drunbar D, Brown BA, Onyi G, Dunlap R, Ahn CH, Murphy DT (1994) Rifampin-resistant *Mycobacterium kansasii*. Clin Infect Dis 18:736–743

Wallace RJ Jr, Brown BA, Griffith DE, Girard WM, Murphy DT (1996) Clarithromycin regimens for pulmonary *Mycobacterium avium* complex. The first 50 patients. Am J Respir Crit Care Med 153:1766–1772

Wallace RJ Jr, Brown-Elliott BA, McNulty S, Philley JV, Killingley J, Wilson RW, York DS, Shepherd S, Griffith DE (2014) Macrolide/azalide therapy for nodular/bronchiectatic *Mycobacterium avium* complex lung disease. Chest 146:276–282

Winthrop KL, Chang E, Yamashita S, Iademarco MF, LoBue PA (2009) Non-tuberculous mycobacteria infections and anti-tumor necrosis-α therapy. Emerg Infect Dis 15:1556–1561

Winthrop KL, McNelley E, Kendall B, Marshall-Olson A, Morris C, Cassidy M, Saulson A, Hedberg K (2010) Pulmonary nontuberculous mycobacterial disease prevalence and clinical features: an emerging public health disease. Am J Respir Crit Care Med 182:977–982

Advs Exp. Medicine, Biology - Neuroscience and Respiration (2017) 27: 27–33
DOI 10.1007/5584_2016_46

Published online: 12 July 2016

Diagnosis of Invasive Pulmonary Aspergillosis

E. Swoboda-Kopeć, M. Sikora, K. Piskorska, M. Gołaś, I. Netsvyetayeva, D. Przybyłowska, and E. Mierzwińska-Nastalska

Abstract

Culturing strains from clinical samples is the main method to diagnose invasive pulmonary aspergillosis. Detecting the galactomannan antigen in serum samples is an auxiliary examination. The goal of this study was to determine the frequency with which *Aspergillus fumigatus* was cultured in clinical samples taken from patients hospitalized in the the Infant Jesus Teaching Hospital in Warsaw, Poland, in the period of 2013–2014. Specimens from the respiratory tract and blood were cultured for mycological and serological assessments. Strain isolation was performed in chloramphenicol Sabouraud agar. Species identification was based on morphological traits in macro-cultures and on microscopic examination. The galactomannan antigen was detected by ELISA method. Out of 2000 clinical samples with positive mycological results, 200 were obtained from the respiratory tract. *A. fumigatus* was cultured in 13 cases from the respiratory group. Ten cases were cultured out of tracheal aspirates and three from bronchoalveolar lavage fluid. The galactomannan antigen was detected in a serum sample from only one out of the 13 patients with cultures positive for *A. fumigatus*. It also was detected in serum samples of three other patients in whom *A. fumigatus* culture yielded a negative result. We conclude that culture-confirmed invasive pulmonary aspergillosis represents a scarce finding. *A. fumigatus* cultured from clinical samples may not always be confirmed by ELISA assay and *vice versa* a positive ELISA result does not attest the successful culture.

The original version of this chapter has been revised. An erratum to this chapter can be found at DOI 10.1007/978-3-319-44488-8_11

E. Swoboda-Kopeć, K. Piskorska (✉), M. Gołaś, and I. Netsvyetayeva
Department of Medical Microbiology, Medical University of Warsaw, 5 Chałubińskiego St., 02-004 Warsaw, Poland
e-mail: kaspiskorska@gmail.com

M. Sikora
Department of Dental Microbiology, Medical University of Warsaw, Warsaw, Poland

D. Przybyłowska and E. Mierzwińska-Nastalska
Department of Prosthodontic, Medical University of Warsaw, Warsaw, Poland

Keywords
Aspergillosis • *Aspergillus fumigates* • Diagnostics • ELISA • Galactomannan antigen • Lungs • Respiratory tract

1 Introduction

A delay in the diagnosis of fungal infections is a consequence of unspecific clinical symptoms and low sensitivity and specificity of the available diagnostic tests (Haddad et al. 2015). Invasive aspergillosis (IA) is an acute, most often fatal, rapidly progressing infection in immunodeficient patients. It usually prevails in the respiratory tract, but may spread to other tissues and organs (Kousha et al. 2011). *Aspergillus fumigates* is responsible for most infections; other species less frequently lead to pulmonary aspergillosis (Chabi et al. 2015; Maturu and Agarwal 2015). Risk factors for invasive aspergillosis include prolonged neutropenia (<500 cells $\cdot$ mm^{-3} for >10 days), transplantation (highest risk for lung transplants and hematopoietic stem cell transplantation), prolonged (>3 weeks) high-dose corticosteroid therapy, hematological malignancy (higher risk in leukemia), chemotherapy, advanced acquired immune deficiency syndrome (AIDS), and chronic granulomatous disease (Kousha et al. 2011).

Standard diagnostics of aspergillosis include CT scans, biopsies, microscopic imaging, and cultures from tissue specimens (samples of sterile body fluids and tissue sections) (Santos et al. 2015). Serological tests, such as the ELISA assay, often used to detect the components of the fungal cell wall, are frequently performed as a complementary diagnostic test (Haddad et al. 2015). They detect galactomannan, the main component of the *Aspergillus spp.* cell wall, released during hypha growth. This antigen may be detected in both serum and bronchoalveolar lavage fluid (BALF). However, BALF is more sensitive and specific than the serum (Haddad et al. 2015; Kousha et al. 2011; Wheat and Walsh 2008). False negative/positive results are still a problem (Marr et al. 2005; Singh et al. 2004). Therefore, it is crucial to thoroughly check the patient's medical history and monitor their condition to be able to assess whether the results are false or not.

Molecular diagnostics consisting of BALF and serum analyses by PCR is an alternative way of diagnosing invasive pulmonary aspergillosis. Detection in serum is more sensitive and specific (100 % and 65–92 %, respectively) than in BALF (67–100 % and 55–95 %, respectively) (Halliday et al. 2006; Hizel et al. 2004). However, the results may be false positive, as there always is risk of inhaling fungal spores. Furthermore, molecular diagnostics do not differentiate between colonization and infection.

Standard diagnostics of fungal infections includes clinical observation and laboratory tests. The definition of fungal infection was published by the European Organization for Research and Treatment of Cancer/Invasive Fungal Infections Cooperative Croup and the National Institute of Allergy and Infectious Diseases Mycoses Study Group (EORTC/MSG) in 2002. It has been introduced to simplify the identification of similar patients for clinical trials and epidemiological studies (de Pauw et al. 2008; Ascioglu et al. 2002).

According to EORTC/MSG guidelines, there are three levels of probability of invasive fungal infections: proven, probable, and possible. The diagnosis of a possible fungal infection is based only on clinical symptoms and risk factors in a patient. Probable and proven infections are additionally classified according to a positive sample culture result and/or the detection of antigen in blood serum (Haddad et al. 2015). Figure 1 presents the details of the classification of invasive infections according to EORTC/MSG guidelines.

Fig. 1 Classification of invasive infections according to EORTC/MSG guidelines (de Pauw et al. 2008; Ascioglu et al. 2002)

2 Methods

The Bioethical Commission of the Medical University of Warsaw in Poland waived the requirement for ethical approval due to the retrospective nature of this research. The prevalence of *Aspergillus spp.* infection in patients hospitalized in the the Infant Jesus Teaching Hospital in Warsaw, Poland during the years 2013–2014 was retrospectively investigated. Routine mycological tests revealed 200 isolates of *Aspergillus spp*. from the respiratory tract, including 184 tracheal aspirates and 16 bronchoalveolar lavages, among 2000 clinical samples of fungi.

2.1 Culture and Identification

Strains were cultured on the Sabouraud agar with chloramphenicol at 30 °C for 24–48 h and were then identified by morphological characteristics under a microscope. Microscopic features of *A. fumigatus* consisted of typical columnar, uniseriate conidial heads. Conidiophores are short, smooth-walled, and have conical shaped terminal vesicles which support a single row of phialides on the upper two thirds of the vesicle. Conidia are produced in basipetal succession, forming long chains. Conidia are globose to subglobose, green, and rough-walled to echinulate (Andreoni et al. 2004).

2.2 Galactomannan (GM) Platelia *Aspergillus* Enzyme Immunoassay (EIA)

The serum was taken for serological testing. All collected serum samples were stored at −20 °C until use. The sandwich ELISA method was employed to detect galactomannan in the serum according to the manufacturer's instruction (Platelia Aspergillus protocol; BioRad, Marnes-la-Coquette, France). Optical density was measured spectrophotometrically with BioRad Model PR5100 ELISA microplate reader. The results were interpreted as based on the index calculated from the measured optical density using the 450 nm wavelength. A cut-off index of 0.5 was considered positive.

3 Results

Two thousand clinical samples testing positive for fungi, including 200 from the respiratory tract, were analyzed within a 2-year period of 2013–2014. *A. fumigatus* was identified by culture in 13 patients in specimens taken from the respiratory tract on the basis of typical morphologic traits. Ten specimens came from tracheal aspirates and three from BALF. Yeast-like fungi, predominantly *Candida albicans*, were other most frequently isolated species from other respiratory tract samples. Table 1 presents the species distribution.

Serum samples of the patients in whom fungi were isolated from the respiratory tract were tested for the galactomannan antigen. Out of the 200 samples, the antigen was detected in four taken from patients after liver and kidney transplantation, i.e., being likely immunocompromised. In one of those patients, a culture for *A. fumigatus* from BALF yielded a positive result. In the remaining three, culture confirmation of the presence of *A. fumigatus* failed. The antigen was not detected in the other 12 patients who had positive cultures.

4 Discussion

It is difficult to determine the prevalence of invasive pulmonary aspergillosis. Immunoassay diagnostics, histological and mycological testing, as well as radiological imaging remain imperfect. The use of just one diagnostic method is usually insufficient to confirm the diagnosis. The prevalence of invasive pulmonary aspergillosis may thus be underestimated. Risk groups for invasive aspergillosis should be better defined and new diagnostic methods which could accelerate the diagnosis should be implemented (Backx et al. 2014; Desoubeaux et al. 2014). The current methods are imperfect and the confirmation of infection by microbiological tests delays the diagnosis, which often leads to fulminant infection at the time of successful diagnosis. Further, microbiological tests all too often fail to confirm the diagnosis *in vivo*, and the cause of infection is discovered posthumously (Maertens et al. 2007) (Table 2).

Table 1 Prevalence of fungi isolated from respiratory tract

Species	Strains (n)	%
C. albicans	111	55.5
C. glabrata	26	13.0
C. parapsilosis	9	4.5
C. tropicalis	9	4.5
C. krusei	7	3.5
C. kefyr	8	4.0
Other yeasts and yeast-like fungi	13	6.5
Aspergillus fumigatus	13	6.5
Other molds	4	2.0
Total	200	100

Perfect diagnostic markers for invasive fungal infections are still searched for. Such markers should be suitable for early disease stages, should differentiate colonisation from infection, be etiologically specific, and not cross-react with human or other microbial antigens. Further, a test detecting a given marker should be easy to perform, standardized, and validated (Yeo and Wong 2002).

In the present study, *A. fumigatus* was cultured once in 13 out of the 200 patients at high risk of invasive fungal infection; ten cases were cultured from tracheal aspirates and three from BALF. Only one of the patients with *A. fumigatus* cultured from BALF, was classified as proven invasive aspergillosis according to the clinical and microbiological criteria developed by the EORTC/MSG guidelines (de Pauw et al. 2008). The patient manifested clinical symptoms of a respiratory tract infection and radiological images showed lesions pointing to active respiratory infection. In this patient, aspergillosis was conclusively confirmed by detecting the galactomannan antigen in a serum sample. The remaining 12 patients in whom *A. fumigatus* was positively cultured could not be classified to any of the three EORTC/MSG classes. These patients did not have any clinical symptoms of invasive aspergillosis and no galactomannan antigen was detected in the serum. The failed detection of the antigen in the presence of positive culture results could have to do with the implemented antifungal treatment.

Galactomannan was detected in three other patients in whom *Aspergillus spp.* culture yielded

Table 2 Advantages and disadvantages of different diagnostic methods in invasive pulmonary aspergillosis

Method	Advantages	Disadvantages
Culture	Simple and cheap;	Time-consuming;
	Allows to identify the fungus and to perform antifungal susceptibility testing;	Low sensitivity;
	High rate of isolation in blood cultures for *Fusarium spp.*	Results conditioned by proper sampling;
		Possibility of contamination;
		Recently improved strategy needs testing in multiple laboratories for other molds.
Galactomannan (GM) detection	Non-invasive method;	Differences in diagnostic and prognostic values in non-neutropenic patients;
	Useful for early diagnosis;	Mold antifungal drug therapy influences detection sensitivity;
	Reproducible methodology;	Persistent GM antigenemia during therapy points to poor prognosis and the need to reassess therapy.
	Greater sensitivity and specificity for mold diagnosis, particularly when non-fumigatus *Aspergillus spp.* are involved.	
1,3-β-glucan detection	Non-invasive method;	False −/+ results (bacteremia);
	Useful and reproducible method, particularly in early diagnostics;	Limited experience, less widely used than GM detection;
	Broad coverage of fungal species;	Threshold for positive results depends on a kind of test used;
	Can be used to screen patients with suspected mold disease;	Antigenemia wanes in invasive aspergillosis and pneumocystis pneumonia under antifungal therapy;
	Useful in patients under antifungal therapy.	Antigen may remain for long above usual threshold after disappearance of clinical symptoms of primary infection;
		Less accurate in hematological patients.
Molecular techniques: Real-Time PCR	Non-invasive method;	Non-mycological criterion is still in development;
	Useful and reproducible method, particularly in early diagnostics;	Limited to reference laboratories, thus of low accessibility;
	Broad coverage of fungal species;	High cost, sophisticated equipment;
	Can be used to screen patients with potential mold disease;	Laborious and difficult efficient fungal DNA extraction from untoward clinical specimens.
	Useful in patients under antifungal therapy;	
	Assessment of mold species and molecular susceptibility.	
Imaging	Computer tomography helpful in early diagnosis;	Not accepted as proof of a mold disease;
	X-ray raises attention to the possibility of mold disease when pulmonary symptoms are manifested.	Image specific not only to invasive aspergillosis (halo sign observed also in other lesions);
		X-ray does not show a fungus ball, therefore requires assessment by highly trained staff.
Histology	Can be useful in diagnostics;	Requires invasive sampling;
	Fungal hyphae are rapidly detected in tissue.	Rarely recommended by physicians;
		Unable to determine the species.

a negative result, although they manifested clinical symptoms of respiratory tract infections. These patients were classified as probable aspergillosis.

Okuturlar et al. (2015) have analyzed 165 cases from the invasive fungal infection risk group. Fifteen patients were diagnosed with invasive pulmonary aspergillosis by microbiological testing, including four qualified as proven and 11 as probable infection. *A. fumigatus* was cultured from nine clinical specimens. Sönmez et al. (2015) have described 199 patients at high risk of invasive fungal infection and classified three patients as proven and 16 as probable infection, distinguishing high and low risk of infection. The galactomannan antigen was detected in serum samples of the three patients classified as proven infection.

The diagnostics of invasive aspergillosis, including the detection of the galactomannan antigen has been described in a number of studies for different patient groups (Leeflang et al. 2008; Pfeiffer et al. 2006). Detection of galactomannan apparently is well associated with *Aspergillus spp*. infection (Backx et al. 2014; Cordonnier et al. 2009). Therefore, most clinical studies rely mainly on galactomannan detection to diagnose probable invasive aspergillosis (Ceesay et al. 2015; Marks et al. 2011). According to the EORTC/MSG guidelines, galactomannan detection is the only reliable and recommended test supporting the diagnosis and enabling the classification of patients to the proven, probable, and possible groups. Monitoring the presence of this antigen may be used to implement pre-emptive therapy in patients with clinical symptoms of an infection before onset of a full-blown disease and visible lesions (Backx et al. 2014; Cordonnier et al. 2009; Maertens et al. 2005). However, in the present study, the galactomannan antigen was detected less frequently than *A. fumigatus* was cultured. Similar findings have been reported in some other studies (Sönmez et al. 2015; Wheat and Walsh 2008), which may put the essential diagnostic role of the detection of galactomannan into doubt. Differential diagnostics consisting of complimentary tests such as culture, galactomannan detection, and others narrows down the search for the etiologic background of an infection and distinguishes the risk groups; thus distinctly helping the decision making concerning the implementation of antifungal treatment.

Conflicts of Interest The authors declare no conflicts of interest in relation to this article.

References

Andreoni S, Farina C, Lombardi G (2004) Medical mycology atlas. Gilead Systems. GRAFIK@rt srl – Paderno Dugnano

Ascioglu S, Rex JH, de Pauw B, Bennett JE, Bille J, Crokaert F, Denning DW, Donnelly JP, Edwards JE, Erjavec Z, Fiere D, Lortholary O, Maertens J, Meis JF, Patterson TF, Ritter J, Selleslag D, Shah PM, Stevens DA, Walsh TJ, Invasive Fungal Infections, Cooperative Group of the European Organization for Research and Treatment of Cancer, Mycoses Study Group of the National Institute of Allergy and Infectious Diseases (2002) Defining opportunistic invasive fungal infections in Immunocompromised patients with cancer and hematopoietic stem cell transplants: an international consensus. Clin Infect Dis 34:7–14

Backx M, White PL, Barnes RA (2014) New fungal diagnostics. Br J Hosp Med (Lond) 75:271–276

Ceesay MM, Desai SR, Berry L, Cleverley J, Kibbler CC, Pomplun S, Nicholson AG, Douiri A, Wade J, Smith M, Mufti GJ, Pagliuca A (2015) A comprehensive diagnostic approach using galactomannan, targeted β-d-glucan, baseline computerized tomography and biopsy yields a significant burden of invasive fungal disease in at risk haematology patients. Br J Haematol 168:219–229

Chabi ML, Goracci A, Roche N, Paugam A, Lupo A, Revel MP (2015) Pulmonary aspergillosis. Diagn Interv Imaging 96:435–442

Cordonnier C, Botterel F, Ben Amor R, Pautas C, Maury S, Kuentz M, Hicheri Y, Bastuji-Garin S, Bretagne S (2009) Correlation between galactomannan antigen levels in serum and neutrophil counts in haematological patients with invasive aspergillosis. Clin Microbiol Infect 15:81–86

De Pauw B, Walsh TJ, Donnelly JP, Stevens DA, Edwards JE, Calandra T, Pappas PG, Maertens J, Lortholary O, Kauffman CA, Denning DW, Patterson TF, Maschmeyer G, Bille J, Dismukes WE, Herbrecht R, Hope WW, Kibbler CC, Kullberg BJ, Marr KA, Muñoz P, Odds FC, Perfect JR, Restrepo A, Ruhnke M, Segal BH, Sobel JD, Sorrell TC, Viscoli C, Wingard JR, Zaoutis T, Bennett JE (2008) Revised definitions of invasive fungal disease from the European Organization for Research and Treatment of Cancer/Invasive Fungal Infections Cooperative Group and the National Institute of Allergy and Infectious Diseases Mycoses Study Group (EORTC/MSG) Consensus Group. Clin Infect Dis 46:1813–1821

Desoubeaux G, Bailly É, Chandenier J (2014) Diagnosis of invasive pulmonary aspergillosis: updates and recommendations. Med Mal Infect 44:89–101

Haddad TM, Narayanan MA, Shaw KE, Vivekanandan R (2015) Fatal case of probable invasive aspergillosis after five years of heart transplant: A case report and review of the literature. Case Rep Infect Dis 2015:864545. doi:10.1155/2015/864545

Halliday C, Hoile R, Sorrell T, James G, Yadav S, Shaw P, Bleakley M, Bradstock K, Chen S (2006) Role of prospective screening of blood for invasive aspergillosis by polymerase chain reaction in febrile neutropenic recipients of haematopoietic stem cell transplants and patients with acute leukaemia. Br J Haematol 132:478–486

Hizel K, Kokturk N, Kalkanci A, Ozturk C, Kustimur S, Tufan M (2004) Polymerase chain reaction in the diagnosis of invasive aspergillosis. Mycoses 47:338–342

Kousha M, Tadi R, Soubani AO (2011) Pulmonary aspergillosis: a clinical review. Eur Respir Rev 20:156–174

Leeflang MM, Debets-Ossenkopp Y, Visser CE, Scholten RJ, Hooft L, Bijlmer HA, Reitsma JB, Bossuyt PM, Vandenbroucke-Grauls CM (2008) Galactomannan detection for invasive aspergillosis in immunocompromized patients. Cochrane Database Syst Rev 4: CD007394. doi:10.1002/14651858.CD007394

Maertens J, Theunissen K, Verhoef G, Verschakelen J, Lagrou K, Verbeken E, Wilmer A, Verhaegen J, Boogaerts M, Van Eldere J (2005) Galactomannan and computed tomography-based preemptive antifungal therapy in neutropenic patients at high risk for invasive fungal infection: a prospective feasibility study. Clin Infect Dis 41:1242–1250

Maertens J, Theunissen K, Lodewyck T, Lagrou K, Van Eldere J (2007) Advances in the serological diagnosis of invasive Aspergillus infections in patients with haematological disorders. Mycoses Suppl 1:2–17

Marks DI, Pagliuca A, Kibbler CC, Glasmacher A, Heussel CP, Kantecki M, Miller PJS, Ribaud P, Schlamm HT, Solano C, Cook G, IMPROVIT Study Group (2011) Voriconazole versus itraconazole for antifungal prophylaxis following allogeneic haematopoietic stem-cell transplantation. Br J Haematol 155:318–327

Marr KA, Laverdiere M, Gugel A, Leisenring W (2005) Antifungal therapy decreases sensitivity of the Aspergillus galactomannan enzyme immunoassay. Clin Infect Dis 40:1762–1769

Maturu VN, Agarwal R (2015) Acute invasive pulmonary aspergillosis complicating allergic bronchopulmonary aspergillosis: case report and systematic review. Mycopathologia 180:209–215

Okuturlar Y, Ozkalemkas F, Ener B, Serin SO, Kazak E, Ozcelik T, Ozkocaman V, Ozkan HA, Akalin H, Gunaldi M, Ali R (2015) Serum galactomannan levels in the diagnosis of invasive aspergillosis. Korean J Intern Med 30:899–905

Pfeiffer CD, Fine JP, Safdar N (2006) Diagnosis of invasive aspergillosis using a galactomannan assay: a meta-analysis. Clin Infect Dis 42:1417–1427

Santos T, Aguiar B, Santos L, Romaozinho C, Tome R, Macario F, Alves R, Campos M, Mota A (2015) Invasive fungal infections after kidney transplantation: a single-center experience. Transplant Proc 47:971–975

Singh N, Obman A, Husain S, Aspinall S, Mietzner S, Stout JE (2004) Reactivity of platelia Aspergillus galactomannan antigen with piperacillin-tazobactam: clinical implications based on achievable concentrations in serum. Antimicrob Agents Chemother 48:1989–1992

Sönmez A, Eksi F, Pehlivan M, Haydaroglu Sahin H (2015) Investigating the presence of fungal agents in febrile neutropenic patients using different microbiological, serological, and molecular methods. Bosn J Basic Med Sci 15:40–47

Wheat LJ, Walsh TJ (2008) Diagnosis of invasive aspergillosis by galactomannan antigenemia detection using an enzyme immunoassay. Eur J Clin Microbiol Infect Dis 27:245–251

Yeo SF, Wong B (2002) Current status of nonculture methods for diagnosis of invasive fungal infections. Clin Microbiol Rev 15:465–468

Advs Exp. Medicine, Biology - Neuroscience and Respiration (2017) 27: 35–45
DOI 10.1007/5584_2016_47

Published online: 13 July 2016

Rarity of Mixed Species Malaria with *Plasmodium falciparum* and *Plasmodium malariae* in Travelers to Saarland in Germany

Josef Yayan and Kurt Rasche

Abstract

Malaria is an acute, life-threatening infectious disease that spreads in tropical and subtropical regions. Malaria is mainly brought over to Germany by travelers, so the disease can be overlooked due to its nonspecific symptoms and a lack of experience of attending physicians. The aim of this study was to analyze, retrospectively, epidemiological and clinical data from patients examined for malaria. Patient data were collected from hospital charts at the Department of Internal Medicine, Saarland University Medical Center, Germany, for the period of 2004–2012. The data of patients with and without malaria were compared in terms of their epidemiological, demographic, clinical, and medical treatment aspects. We identified found 15 patients with malaria (28.3 %, mean age 42.3 ± 16.5 years, three females [20 %]; 95 % confidence interval of 0.2–0.4) out of the 53 patients examined. Mainly locals brought malaria over to Homburg, Germany ($p = 0.009$). Malaria tropica was the most common species ($p < 0.0001$). One patient (6.7 %) with malaria, who had recently traveled, had a mixed infection of *Plasmodium falciparum* and *Plasmodium malariae* ($p = 0.670$). Malaria is characterized by thrombocytopenia ($p = 0.047$) and elevated C-reactive protein ($p = 0.019$) in serum, and fever is the leading symptom ($p = 0.031$). In most cases, malaria was brought from Ghana (33.3 %). Further, patients had contracted malaria despite malaria prophylaxis

The original version of this chapter has been revised. An erratum to this chapter can be found at DOI 10.1007/978-3-319-44488-8_11

J. Yayan (✉)
Department of Internal Medicine, Division of Pulmonary, Allergy, and Sleep Medicine, Saarland University Medical Center, Kirrberger Straße, Homburg/Saar, Germany

Department of Internal Medicine, Division of Pulmonary, Allergy and Sleep Medicine, HELIOS Clinic Wuppertal, Witten/Herdecke University, Witten, Germany
e-mail: josef.yayan@hotmail.com

K. Rasche
Department of Internal Medicine, Division of Pulmonary, Allergy and Sleep Medicine, HELIOS Clinic Wuppertal, Witten/Herdecke University, Witten, Germany

(33.3 %, p = 0.670). In conclusion, malaria test should be used in patients with fever after a journey from Africa. Malaria caused by *Plasmodium falciparum* is the most common species of brought over malaria. Mixed-species *Plasmodium falciparum* and *Plasmodium malariae* are uncommon in travelers with malaria.

Keywords

Malaria parasites • Malaria symptoms • Malaria treatment • Mosquito • Prophylaxis • Traveler malaria • Tropical disease

1 Introduction

Malaria is an acute infectious disease that can become life threatening within a short time (WHO 2016; Burchard 2011; Gyorkos et al. 1995). It is considered a tropical disease. Malaria is caused by *Plasmodium*. There are four known human pathogenic *Plasmodium* species: *Plasmodium malariae*, *Plasmodium falciparum*, *Plasmodium vivax*, and *Plasmodium ovale*. Malaria parasites are transmitted through the bite of *Anopheles* mosquito. Malaria can be either benign or malignant. The benign form is known as quartan malaria and tertian malaria, and the malignant form is malaria tropica (Lalloo et al. 2007). *Plasmodium malariae* causes malaria quartan, *Plasmodium vivax* and *ovale* cause malaria tertian, and *Plasmodium falciparum* causes malaria tropica. *Plasmodium falciparum* infects a large number of red blood cells and quickly advances to a severe or life-threatening multi-organ disease. Mixed infections with more than one species of parasite can occur; such infections commonly involve *Plasmodium falciparum* with the attendant risk of severe malaria (Lalloo et al. 2007).

Malaria is mostly brought over to Europe by travelers or immigrants from tropical and subtropical geographic regions (Burchard 2011). Epidemiological evidence shows that the origin of infection is correlated with patterns of migration in European countries, i.e., many cases are travelers who have returned to their country of origin. Because the clinical symptoms of malaria are nonspecific and can be difficult to distinguish from a variety of other febrile diseases, malaria must be considered in patients with a fever of unknown origin and a travel history (Burchard 2011). As malaria is not often seen in Europe, its underrecognition by physicians is likely. Underrecognition or misdiagnosis of malaria may also have to do with the fact that it is qute often contracted by travelers residing for a short time in an endemic region, which decreases physicians' attentiveness to this possibility (Kalinowska-Nowak et al. 2012). Clinicians working in highly urbanized areas with significant immigrant communities may see higher numbers of cases of malaria. However, as tourism and globalization are increasing worldwide, it is possible that the number of cases of malaria will increase. Chemoprophylaxis could be helpful in lowering the malaria risk, and it forms one of the key components of prevention of bringing malaria over. According to the World Health Organization's (WHO 2010) ABCD approach to malaria prevention, travelers should: have *a*wareness of risk, avoid being *b*itten by mosquitoes, take *c*hemoprophylaxis correctly, and promptly search for *d*iagnosis and treatment. Travel agents and health practitioners should offer sufficient information about chemoprophylaxis to all travelers to malaria-endemic areas (Gyorkos et al. 1995). This presupposes that the provision of correct information will be effective, but this is not always the case. There are various reasons for the non-uptake of chemoprophylaxis, and travelers' receiving incorrect information is just one possibility (Behrens and Alexander 2013; Morgan and Figueroa-Muñoz 2005).

The present investigation was conducted to analyze the time trends of malaria brought over to the German town of Homburg. Homburg is located in the province of Saarland in western Germany, with a population of 44,000. The University of Saarland Medical Center is the main reference place for infectious diseases in the area and it treats patients with malaria. This investigation also examined the variation in clinical symptoms of patients with malaria, as well as the treatments and outcomes.

2 Methods

All of patients' data were anonymized prior to analysis. The Medical Association of Saarland's Institutional Review Board approved this study. The requirement for written, informed consent of patients was waived because of the retrospective nature of the analysis of medical records.

2.1 Patients

This unmatched case-control study retrospectively examined the whole data from all patients who had been tested for malaria, using hospital chart data from the Department of Internal Medicine, Saarland University Medical Center, during the period of 2004–2012. The data included age, gender, race, duration of hospitalization, outpatient or inpatient status, laboratory values, discharge diagnosis from the local hospital, and current and previous diseases from questioning the personal history of the patient.

2.2 Case-Control Study

Malaria cases and controls were identified by a process of passive case detection in which outpatients with febrile illness were examined for malaria. The case definition included those with clinical symptoms of malaria, such as a body temperature > 37 °C, history of fever, headache, or body ache, who tested positive for malaria parasites by microscopy. Individuals with febrile illness shown by microscopy to be malaria-negative were identified as controls.

The diagnosis of co-morbidities was also carried from the hospital charts, medical history, or during the current investigation newly discovered diseases. Comorbidities were diagnosed after clinical symptoms and confirmed by instrumental and laboratorial examinations. The process of data collection was the same for cases and controls. The selection of cases and controls was made independently using hospital chart data, and individual matching was not undertaken.

2.3 Inclusion and Exclusion Criteria

The inclusion criteria were the following: patients with unclear fever who had recently returned from a malaria-endemic area or any patient whose doctor considered malaria as a differential diagnosis. The study population consisted of patients diagnosed with malaria after malaria parasites or antigens were identified in the patient's blood. The control group consisted of patients in whom malaria was excluded. Age and sex differences were compared between patients with and without malaria. This study examined whether malaria was more common in the local German population or in foreigners. Additionally, as malaria is known to be a tropical and subtropical infectious disease, the travel destination as a possible source of infection was investigated (Behrens and Alexander 2013).

All patients under 18 years of age who were detected to have malaria were excluded from the study. Patients examined at the Department of Neurology who were suspected of having malaria were also excluded from this study because of restricted access to their medical records.

2.4 Symptoms and Diagnosis of Malaria

Early symptoms of malaria are nonspecific and diverse, such as fever, headache, weakness, myalgia, chills, dizziness, abdominal pain, diarrhea, nausea, vomiting, anorexia, and pruritus

(Looareesuwan 1999). These symptoms were compared between patients with and without malaria in the study population. Malaria was diagnosed microscopically by staining thick and thin blood films on a glass slide to visually identify the malaria parasites. Following skin disinfection (74.1 % ethanol and propan-2-ol 10 %) with aerosol Softasept® N (B. Braun Melsungen PLC, Melsungen, Germany), a minimum of 2.7 mL blood in EDTA Monovette® (SARSTEDT, Nümbrecht, Germany) (red top) was taken through venipuncture with a blood-collection needle (Safety-Multifly®, SARSTEDT). A thick blood film was prepared from each patient. The blood spot was smeared using a circular motion with the corner of a spreader slide, being careful not to make the preparation too thick, and permitting the smear to dehydrate without fixing. After drying, the spot was sullied with diluted Giemsa (1:20, vol/vol) for 20 min and washed by placing the film in buffered water for 3 min. The glass slide was allowed to air dry in a vertical position and was examined using a light microscope. As the specimens were unfixed, the red cells lysed when a water-based stain was applied. A thin blood film was prepared by immediately placing the smooth edge of a spreader slide in a drop of blood, adjusting the angle between the slide and spreader to 45°, and then smearing the blood with a swift and steady sweep along the surface. The film was then permitted to air dry and was fixed with complete methanol. After ventilation, the sample was stained with thinned Giemsa (1:20, vol/vol) for 20 min and then washed by briefly plunging the glass slide in and out of a jar of buffered water, as overdone washing would decolorize the film. The glass slide was then allowed to air dry in a vertical position and was inspected under a light microscope (Chotivanich et al. 2007).

The diagnosis of malaria was made according to the gold standard blood smear. The polymerase chain reaction (PCR) technique was used to confirm malaria infection, to monitor follow-up therapeutic response, and to detect drug resistance. The classification of malaria was specified in each case according to the latest edition of the International Classification of Disease (ICD B50-B54). The use of chemoprophylaxis was compared in both groups.

2.5 Laboratory Evaluation

After the sample collection, the amount of C-reactive protein (CRP) in the human serum and plasma was measured in lithium heparin SARSTEDT Monovette® 4.7 mL (orange top) using a standard immuno-turbidimetric assay on the COBAS® INTEGRA system (normal value < 6 mg/L). Simultaneously, total bilirubin (normal range 0.1–1.0 mg/dL), creatinine (normal range 0.7–1.2 mg/dL), and glucose (normal range 55–110 mg/dL) were measured in the plasma. Blood leukocyte count (normal range 4000–10,000/μL) was carried out by flow cytometry after collection in EDTA Monovette® 2.7 mL. Along with the leukocyte count, hemoglobin (normal range 14–18 g/dL), hematocrit (normal range 41–53 %), and platelet (normal range 140,000–400,000/μL) were determined for all patients simultaneously. Thrombocytopenia was defined as a platelet count under 140,000/μL.

2.6 Virological Evaluation

Virological-serological tests were done on the serum of all of the patients to exclude hepatitis infection. For this purpose, two SARSTEDT serum Monovette® 4.7 mL (brown top) and two EDTA Monovette® 2.7 mL were used after a venipuncture. The initial round of examination consisted of hepatitis B serology using enzyme-linked immunosorbent assay (ELISA), hepatitis B surface antigen (HBsAg, anti-HBsAg, and anti-HBcAg), antibodies against hepatitis C virus (including a serological confirmatory test if screening was positive), antibodies against hepatitis A virus immunoglobulin (IgG and IgM), antibodies against cytomegalovirus (CMV) (IgG and IgM); and antibodies against Epstein-Barr-Virus (EBV) (VCA IgG and IgM). The subsequent investigation involved qualitative PCR for hepatitis C, qualitative PCR for hepatitis B if a result of quantification HBeAg and anti-HBeAg was positive, or antibodies against hepatitis E if HBV or HCV PCR was negative.

Next, written informed consent was obtained from all participants to perform a test for the detection of human immunodeficiency virus (HIV). Antibodies against HIV-1, HIV-2, and p24 antigen were investigated by ELISA test in the serum collected with Monovette® 4.7 mL (brown top).

2.7 Statistical Elaboration

Categorical data are expressed in proportion, while continuous data are expressed as means standard deviations (SDs). Calculations were performed at a 95 % confidence interval (CI) for the total number of patients with malaria and for the comparison of the use and non-use of antimalarial chemoprophylaxis. A chi-squared test for two independent variables of two probabilities was carried out for gender difference. This test was also used for statistical comparison of symptoms; locals and foreigners; outpatients and inpatients; and comorbidities between the patients with and without malaria. One-way analysis of variance (ANOVA) for independent samples was performed to compare mean age, duration of hospital stay, and laboratory values between the two groups. The null hypothesis in the study was that there would be no differences for the type of malaria, antimalarial medication, and time trends of malaria. The tests were two-tailed, and a p-value of < 0.05 was considered statistically significant.

3 Results

In the hospital database, 53 patients were found who had been examined for malaria during the study period of 2004–2012. In total, 15 patients (28.3 %, three females [20 %]; 95 % CI, 0.2–0.4) were diagnosed with malaria. There were no differences in gender and age between the patients with and without malaria (Table 1). The patients with malaria were treated as inpatients, and those without malaria were treated as outpatients ($p = 0.01$). There was no difference in the duration of hospitalization (Table 1). Notably, there was a significant increase in malaria cases in 2009 ($p = 0.031$). However, a continued increase in malaria prevalence was not observed during the study period (Fig. 1). Malaria caused by *Plasmodium falciparum* was the most common form of the disease noted in Homburg and the surrounding province of Saarland ($p < 0.0001$) (Table 2). One patient (6.7 %) had a mixed malaria infection from two species of malaria parasite, *Plasmodium falciparum* and *Plasmodium malariae*, in the blood smear and PCR diagnosis ($p = 0.670$). Thrombocytopenia and enhanced level of CRP were the dominant laboratory findings (Table 3), and fever was the leading clinical symptom in the patients with malaria ($p = 0.031$) (Table 4), Malaria was mainly brought over by locals to Homburg from long haul destinations ($p = 0.009$), mainly from Ghana. In contrast, no malaria cases were

Table 1 Basic demographic and hospital data of patients with and without malaria

	Malaria (n = 15) (%)	No malaria (n = 38) (%)	p-value
Male	12 (80.0)	30 (79.0)	0.932
Female	3 (20.0)	8 (21.1)	0.932
Locals	8 (53.3)	33 (86.8)	**0.009**
Foreigners	7 (46.7)	5 (13.2)	**0.009**
Mean age (year) ± SD	42.3 ± 16.5	44.5 ± 18.6	0.691
Length of hospital stay (days) ± SD	6.3 ± 8.3	3.3 ± 9.4	0.291
Outpatients	4 (26.7)	25 (65.8)	**0.010**
Inpatients	11 (73.3)	13 (34.2)	**0.010**

SD standard deviation

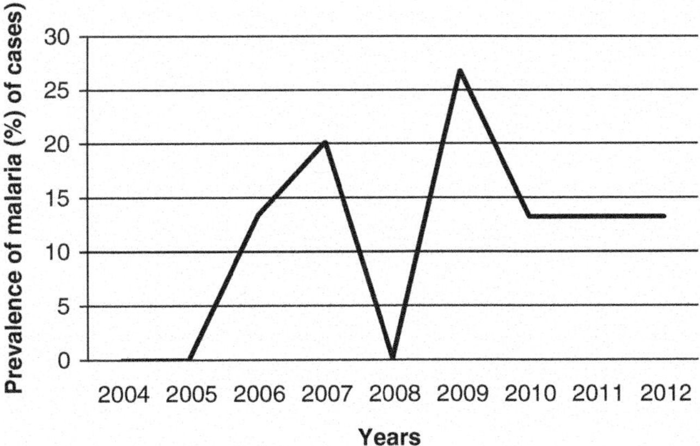

Fig. 1 Time trends in the prevalence of imported malaria to the province of Saarland, Germany in the period of 2004–2012. Percentage of malaria cases in consecutive year out of the total of 15 cases

Table 2 Types and species of malaria brought over from long haul destinations to Homburg, Germany, from 2004 to 2012

Type of malaria	Patients with malaria (n = 15) (%)	p-value	95 % CI
Malaria tropica	11 (73.3)	**<0.0001**	0.50–0.90
Quartan malaria	2 (13.3)	0.667	0.04–0.40
Tertian malaria	2 (13.3)	0.667	0.04–0.40
Plasmodium falciparum	11 (73.3)	**<0.0001**	0.50–0.90
Plasmodium vivax	0	0.197	0.00–0.20
Plasmodium ovale	2 (13.3)	0.667	0.04–0.40
Plasmodium malariae	1 (6.7)	0.670	0.01–0.30
Mixed-species			
Plasmodium falciparum + *Plasmodium malariae*	1 (6.7)	0.670	0.01–0.30

CI confidence interval

Table 3 Comparison of laboratory values between patients with and without malaria

	Malaria (n = 15)	No malaria (n = 38)	p-value
Hemoglobin (14–18 g/dL) ± SD	13.0 ± 4.2	17.7 ± 20.3	0.400
Hematocrit (41–53 %) ± SD	38.3 ± 12.4	41.3 ± 6.0	0.247
Leukocyte count (4000–10,000/μL) ± SD	5478 ± 2056	6434 ± 2301	0.179
Platelet (140,000–400,000/μL) ± SD	137,571 ± 101,143	189,684 ± 73,953	**0.047**
CRP (<6.0 mg/L) ± SD	76.4 ± 75.9	34.1 ± 44.6	**0.019**
Creatinine (0.7–1.2 mg/dL) ± SD	0.8 ± 0.3	1.1 ± 0.7	0.263
Glucose (55–110 mg/dL) ± SD	110 ± 46	103 ± 33	0.525
Bilirubin (0.1–1.0 mg/dL) ± SD	1.7 ± 2.2	1.5 ± 4.2	0.863

Data are means ± SD; *CRP* C-reactive protein

reported from Thailand (Table 5). Patients without malaria had more comorbidities (Table 6). Most of the patients with malaria used the antimalarials atovaquone/proguanil and mefloquine (Table 7). Five male patients (33.3 %, 95 % CI, 0.1–0.6) with malaria and four patients (10.5 %, 95 % CI, 0.008–0.2; one female) without malaria had used antimalarial chemoprophylaxis. Two of those patients (13.3 %, 95 % CI, 0.0–0.3) had discontinued malaria prophylaxis prematurely due to intolerance. There were no deaths in either group.

Table 4 Symptoms in patients with and without malaria

	Malaria (n = 15) (%)	No malaria (n = 38) (%)	p-value
Fever	12 (80.0)	18 (47.4)	**0.031**
Headache	3 (20.0)	9 (23.7)	0.550
Cough	1 (6.7)	3 (7.9)	0.879
Sickness	3 (20.0)	7 (18.4)	0.895
Vomit	3 (20.0)	2 (5.3)	0.098
Diarrhea	3 (20.0)	8 (21.1)	0.932
Chills	6 (40.0)	6 (15.8)	0.058
Body ache	2 (13.3)	5 (13.2)	0.986
Dizziness	2 (13.3)	1 (2.6)	0.129
Weakness	1 (6.7)	0	0.108
Sweats	1 (6.7)	5 (13.2)	0.502
Fatigue	6 (40.0)	8 (21.1)	0.159
Custom cramps	1 (6.7)	0	0.108
Night sweats	1 (6.7)	0	0.108
Dark urine	1 (6.7)	0	0.108
Weight loss	1 (6.7)	0	0.108
Cold sweats	1 (6.7)	0	0.108
Lymph node swelling	1 (6.7)	0	0.108
Reduced general condition	1 (6.7)	0	0.108
Impaired consciousness	1 (6.7)	0	0.108
Brown sputum	0	2 (5.3)	0.365
Pollakiuria	0	1 (2.6)	0.526
Shortness of breath	0	1 (2.6)	0.526
Sore throat	0	2 (5.3)	0.365

Table 5 Geographic origin of malaria in patients who brought it over from faraway destinations and in those in who malaria was excluded

	Malaria (n = 15) (%)	No malaria (n = 38) (%)
Peru	0	1 (2.6)
Ghana	5 (33.3)	2 (5.3)
Thailand	0	6 (15.8)
Togo	2 (13.3)	0
Tanzania	2 (13.3)	0
Congo	1 (6.7)	1 (2.6)
Benin	2 (13.3)	0
Aswan	1 (6.7)	0
Nigeria	1 (6.7)	1 (2.6)
Indonesia	1 (6.7)	0
Brazil	0	1 (2.6)
Dominican Republic	0	1 (2.6)
Asia	0	1 (2.6)
Kenya	0	3 (7.9)
India	0	2 (5.3)
Ecuador	0	1 (2.6)
Egypt	0	2 (5.3)
Other	0	16 (42.1)

Table 6 Comorbidities in patients with and without malaria

	Malaria (n = 15) (%)	No malaria (n = 38) (%)	p-value
Acute bronchitis	0	1 (2.6)	0.526
Acute respiratory distress syndrome	1 (6.7)	0	0.108
Acute tonsillitis	0	1 (2.6)	0.526
Acute urinary tract infection	0	1 (2.6)	0.526
Appendectomy	0	4 (10.5)	0.191
Benign prostatic hyperplasia	0	1 (2.6)	0.526
Cataract	0	1 (2.6)	0.526
Chronic bronchitis	0	1 (2.6)	0.526
Chronic osteomyelitis	0	1 (2.6)	0.526
Chronic polyarthritis	0	1 (2.6)	0.526
Chronic renal failure	0	1 (2.6)	0.526
Condition after brain tumor	0	1 (2.6)	0.526
Condition after stroke	0	1 (2.6)	0.526
Dehydration	1 (6.7)	0	0.108
Diabetes	0	2 (5.3)	0.365
Disseminated intravascular coagulation	1 (6.7)	0	0.108
Drug abuse	0	1 (2.6)	0.526
Early abortion	1 (6.7)	0	0.108
Erysipelas	0	1 (2.6)	0.526
Fecal impaction	0	1 (2.6)	0.526
Fatty liver	0	1 (2.6)	0.526
Hemolytic anemia	0	1 (2.6)	0.526
Hemorrhoid	1 (6.7)	0	0.1081
Herniated disc	0	1 (2.6)	0.526
Hypertension	1 (6.7)	3 (7.9)	0.879
Hyperlipidemia	0	2 (5.3)	0.365
Hypertensive heart disease	0	1 (2.6)	0.526
Hysterectomy	0	1 (2.6)	0.526
Infection of unknown origin	0	2 (5.3)	0.365
Inflammatory response syndrome	1 (6.7)	0	0.108
Influenza	1 (6.7)	8 (21.1)	0.209
Iron deficiency anemia	1 (6.7)	0	0.108
Liver cirrhosis	0	1 (2.6)	0.526
Mycoplasma pneumonia	0	1 (2.6)	0.526
Nicotine	0	3 (7.9)	0.263
Obesity	0	1 (2.6)	0.526
Osteoarthritis	0	1 (2.6)	0.526
Otitis	0	1 (2.6)	0.526
Paracetamol intoxication	1 (6.7)	0	0.108
Pregnancy	1 (6.7)	0	0.108
Reflux esophagitis	1 (6.7)	0	0.108
Sepsis	0	1 (2.6)	0.526
State after hepatitis A	1 (6.7)	0	0.108
State after hepatitis B	1 (6.7)	1 (2.6)	0.487
State after hepatitis C	0	2 (5.3)	0.365
State after amebiasis	0	1 (2.6)	0.526
State after disc surgery	0	1 (2.6)	0.526
State after groin surgery	1 (6.7)	1 (2.6)	0.487
State after herpes zoster	1 (6.7)	0	0.108

(continued)

Table 6 (continued)

	Malaria (n = 15) (%)	No malaria (n = 38) (%)	p-value
State after malaria tropica	0	1 (2.6)	0.526
State after nasal septum surgery	1 (6.7)	0	0.108
State after prostate surgery	0	1 (2.6)	0.526
State after dysentery	0	1 (2.6)	0.526
State after tuberculosis	0	1 (2.6)	0.526
Thrombosis	0	1 (2.6)	0.526

Table 7 Malaria medication treatment and prophylaxis in patients (n) (%)

Antimalarial medication	(n = 15) (%)	p-value	95 % CI
Atovaquone/Proguanil	9 (60.0)	**<0.0001**	0.4–0.8
Mefloquine	4 (26.7)	**0.031**	0.1–0.5
Quinine	2 (13.3)	0.667	0.04–0.4
Malaria prophylaxis			
Chemoprophylaxis	10 (66.7)	0.670	0.01–0.3
No chemoprophylaxis	5 (33.3)	0.670	0.01–0.3

4 Discussion

Malaria is mainly brought over to Homburg by local travelers coming back from tropical and subtropical regions. Most malaria patients in this study were infected with a single *Plasmodium species*. One patient with malaria who had recently traveled was infected by two *Plasmodium* species: *Plasmodium falciparum* and *Plasmodium malariae*. The prevalence of mixed malaria infections was just low, which is in line with other reports on the subject (Mehlotra et al. 2000; Snournou et al. 1993). These findings propose that mixed *Plasmodium* species malaria infections are not the usual occurrence in travelers arriving from regions with malaria epidemics.

In the present study, a continuous increase in the amount of malaria cases was not observed nor was there a clear time trend over the years. The mean age of patients with malaria was 42 years. Most malaria cases came from Africa, specifically from Ghana, and *Plasmodium falciparum* was the most often infection agent encountered, which in line with other studies (Leder et al. 2004). Most malaria patients were German travelers to long haul destinations. We noted no malaria-related deaths. Many patients with malaria did not use pharmacological prophylaxis and some of those who did also contracted the disease, which might be caused by the unfinished course of treatment. An increased serum CRP level and thrombocytopenia were the most significant laboratory features in malaria patients, which is in line with the study of Epelboin et al. (2013). For comparison, the Robert Koch Institute, where the central database for malaria exists for statistical analysis in Germany, has reported that there is no clear time trend in malaria cases from 1993 to 2003 (Schöneberg et al. 2005). However, this institute has reported that the percentage of cases in persons aged 40–49 has, in general, increased over time and the greatest malaria frequency has been those aged 20–39. Most cases of malaria has come from Africa, Asia, Central America, and South America and *Plasmodium falciparum* has been the prevalent infectious species. Further, most malaria patients (60 %) were German travelers who had gone overseas on holidays or to study. Most malaria sufferers did not prophylaxis and more than 20 deaths from malaria were reported in 1999 (Schöneberg et al. 2001).

The most common travel-related diseases are gastrointestinal, febrile, and dermatologic. A fever in returning travelers necessitates immediate attention, as it could be the appearance of a quickly progressing and deadly infection

(Burchard 2014). The diagnostics for the febrile patient should be stepwise and take into account the patient's travel history and experience. Malaria is the most common cause of fever in patients returning from tropical and subtropical areas. This observation was also confirmed in the present study, where most patients diagnosed with malaria presented with fever. However, it is unclear if the most common cause of fever in returning travelers is malaria infection. Severe cases of malaria, with patients staying in the intensive care unit, have poor outcomes (Corne et al. 2004). Such cases of malaria were not observed in the present study, although most patients with malaria were hospitalized. All malaria patients should be hospitalized within 24 h (Lalloo et al. 2007) and treated in a specialized department of a hospital (Marks et al. 2014).

Numerous antimalarial medications exist for chemoprophylaxis in case of traveling to a malaria-endemic country or for treating a patient with malaria. The selection of the most suitable medication should be influenced by contraindications and indications for each drug, principally subordinated to the existence and level of *Plasmodium falciparum* chemosensitivity in the region where the infection originated. For prevention, chloroquine alone can be considered for visitors to regions where *Plasmodium falciparum* is rare or not chloroquine resistant. For other regions, selection of chloroquine and proguanil or mefloquine varies depending on the frequency of *Plasmodium falciparum* and its level of resistance to chloroquine (Bourgeade and Delmont 1998). For treatment, the only indications for chloroquine are malaria cases brought over from faraway destinations due to either *Plasmodium vivax*, *Plasmodium ovale*, or *Plasmodium malariae* or affected by *Plasmodium falciparum* contracted in one of the few nations where this type is still responsive to chloroquine. For non-severe *Plasmodium falciparum* malaria developed in a chemo-resistant zone, mefloquine, halofantrine, sulfadoxine-pyrimethamine, or oral quinine is prescribed, depending on the suitable chemoprophylaxis, contraindications, and doubt of chemo-resistance. Irrespective of the region of origin, *Plasmodium falciparum* in a patient with severe symptoms or heavy vomiting, is treated by intravenous quinine, associated with tetracycline if the patient arrives from a zone that is known for little quinine sensitivity of this species (Bourgeade and Delmont 1998). In the present study, mefloquine was employed for treatment of most malaria patients, and a combination treatment with atovaquone and proguanil was employed in a few most severe cases. The antimalarial medications prescribed most often in another study have consisted of atovaquone and proguanil (Bloechliger et al. 2014). Mefloquine is frequently prescribed for children and pregnant women; however, it is rarely prescribed for patients with comorbidities considered as contraindications. A combination of chloroquine and proguanil is less efficient compared with mefloquine, but that of atovaquone and proguanil is at least as effective as mefloquine, although disease progression may take place (Kofoed and Petersen 2003). However, there are discrepancies in treatment outcomes among studies. Jacquerioz and Croft (2009) have reported that atovaquone-proguanil and doxycycline are the top tolerated medications, and mefloquine is linked with neuropsychiatric adverse outcomes. Siikamäki et al. (2013) have found that mefloquine, atovaquone and proguanil, and doxycycline are efficient as a means of chemoprophylaxis against *Plasmodium falciparum* malaria, when properly employed.

5 Conclusions

Patients with a fever after returning from Africa should be subjected to a malaria test. Malaria caused by *Plasmodium falciparum* is the most common form of disease brought over to Saarland province in Germany by local travelers to Africa and other long haul destinations. Mixed-species *Plasmodium falciparum* and *Plasmodium malariae* are uncommon in malaria in travelers. Fever, thrombocytopenia, and elevated CRP are the leading features of malaria. Malaria prophylaxis is recommended as a common policy for travelers to Africa. Malaria must be excluded in the process of differential diagnosis in all people exhibiting fever who have recently returned from the tropics.

Conflicts of Interest The authors report no conflicts of interest in relation to this article.

References

Behrens RH, Alexander N (2013) Malaria knowledge and utilization of chemoprophylaxis in the UK population and in UK passengers departing to malaria-endemic areas. Malar J 12:461

Bloechliger M, Schlagenhauf P, Toovey S, Schnetzler G, Tatt I, Tomianovic D, Jick SS, Meier CR (2014) Malaria chemoprophylaxis regimens: a descriptive drug utilization study. Travel Med Infect Dis 12(6 Pt B):718–725

Bourgeade A, Delmont J (1998) The proper use of antimalarial drugs currently available. Bull Soc Pathol Exot 91:493–496

Burchard GD (2011) Malaria. Internist (Berl) 55:165–176

Burchard G (2014) Fever in returning travelers. Internist (Berl) 55:274–279

Chotivanich K, Silamut K, Day NPJ (2007) Laboratory diagnosis of malaria infection – a short review of methods. NZ J Med Lab Sci 61:4–7

Corne P, Klouche K, Basset D, Amigues L, Béraud JJ, Jonquet O (2004) Severe imported malaria in adults: a retrospective study of 32 cases admitted to intensive care units. Pathol Biol (Paris) 52:622–626

Epelboin L, Boullé C, Ouar-Epelboin S, Hanf M, Dussart P, Djossou F, Nacher M, Carme B (2013) Discriminating malaria from dengue fever in endemic areas: clinical and biological criteria, prognostic score and utility of the C-reactive protein: a retrospective matched-pair study in French Guiana. PLoS Negl Trop Dis 7:e2420

Gyorkos TW, Svenson JE, Maclean JD, Mohamed N, Remondin MH, Franco ED (1995) Compliance with antimalarial chemoprophylaxis and the subsequent development of malaria: a matched case-control study. Am J Trop Med Hyg 53:511–517

Jacquerioz FA, Croft AM (2009) Drugs for preventing malaria in travellers. Conchrane Database Syst Rev CD006491. doi:10.1002/14651858.CD006491.pub2

Kalinowska-Nowak A, Bociaga-Jasik M, Leśniak M, Mach T, Garlicki A (2012) The risk of malaria during travel, observations in the department of infectious diseases in Cracow from 1996 to 2010. Przegl Epidemiol 66:431–436 (Article in Polish)

Kofoed K, Petersen E (2003) The efficacy of chemoprophylaxis against malaria with chloroquine plus proguanil, mefloquine, and atovaquone plus proguanil in travelers from Denmark. J Travel Med 10:150–154

Lalloo DG, Shingadia D, Pasvol G, Chiodini PL, Whitty CJ, Beeching NJ, Hill DR, Warrell DA, Bannister BA, HPA Advisory Committee on Malaria Prevention in UK Travellers (2007) UK malaria treatment guidelines. J Infect 54:111–121

Leder K, Black J, O'Brien D, Greenwood Z, Kain KC, Schwartz E, Brown G, Torresi J (2004) Malaria in travelers: a review of the GeoSentinel surveillance network. Clin Infect Dis 39:1104–1112

Looareesuwan S (1999) Malaria. In: Looareesuwan S, Wilairatana P (eds) Clinical tropical medicine. Medical Media, Bangkok, pp 5–10

Marks M, Gupta-Wright A, Doherty JF, Singer M, Walker D (2014) Managing malaria in the intensive care unit. Br J Anaesth 113:910–921

Mehlotra RK, Lorry K, Kastens W, Miller SM, Alpers MP, Bockarie M, Kazura JW, Zimmerman PA (2000) Random distribution of mixed species malaria infections in Papua New Guinea. Am J Trop Med Hyg 62:225–231

Morgan M, Figueroa-Muñoz JL (2005) Barriers to uptake and adherence with malaria prophylaxis by the African community in London, England: focus group study. Ethn Health 10:355–372

Schöneberg I, Strobel H, Apitzsch L (2001) Malaria incidence in Germany 1998/99 – results of single case studies of the Robert Koch Institute. Gesundheitswesen 63:319–325

Schöneberg I, Stark K, Attmann D, Krause G (2005) Malaria in Germany 1993 to 2003. Data from the Robert Koch Institute on affected groups of people, countries traveled to and treatment. Dtsch Med Wochenschr 130:937–941

Siikamäki H, Kivelä P, Lyytikäinen O, Kantele A (2013) Imported malaria in Finland 2003–2011: prospective nationwide data with rechecked background information. Malar J 12:93

Snournou G, Viriyakosol S, Jarra W, Thaithong S, Brown KN (1993) Identification of the four human malaria parasite species in field samples by the polymerase chain reaction and detection of a high prevalence of mixed infections. Mol Biochem Parasitol 58:283–292

WHO (2010) Guidelines for the treatment of malaria in travelers. World Health Organization, Geneva. http://www.who.int/malaria/docs/TreatmentGuidelines2010.pdf. Accessed 15 May 2016

WHO (2016) Media Centre Malaria. http://www.who.int/mediacentre/factsheets/fs094/en/. Accessed 15 May 2016

Advs Exp. Medicine, Biology - Neuroscience and Respiration (2017) 27: 47–52
DOI 10.1007/5584_2016_59

Published online: 29 July 2016

Antibiotics Modulate the Ability of Neutrophils to Release Neutrophil Extracellular Traps

A. Manda-Handzlik, W. Bystrzycka, S. Sieczkowska, U. Demkow, and O. Ciepiela

Abstract

Antibiotics directly inhibit the growth and kill microorganisms, and many of them have immunomodulatory properties. We investigated the influence of cefotaxime and gentamicin on the release of neutrophil extracellular traps (NETs) – recently described strategy employed by neutrophils to fight infections. We found that gentamicin inhibits NETs release by human neutrophils, while cefotaxime did not have any impact on this process. The information that antibiotics can modulate NETs release, can be useful in the therapy of infectious diseases in patients suffering from NET-related diseases.

Keywords

Cefotaxime • Gentamicin • Granulocytes • Phagocytosis • NETs

1 Introduction

Antibiotics are widely used for prevention and treatment of bacterial infections. Their antimicrobial effect is not limited to inhibiting the growth and killing microorganisms – they can exert a number of effects on the immune system function. Antibiotics can modulate the patterns of cytokine release (Ziegeler et al. 2006), augment the antigen recognition system (Tentori et al. 1998), and suppress delayed-type hypersensitivity reaction (Roszkowski et al. 1985). Moreover, there is a wide range of interactions between antibiotics and phagocytic cells. Antibiotics inhibit chemotaxis, oxidant production, and affect phagocytic functions (Minic

The original version of this chapter has been revised. An erratum to this chapter can be found at DOI 10.1007/978-3-319-44488-8_11

A. Manda-Handzlik (✉)
Department of Laboratory Diagnostics and Clinical Immunology of Developmental Age, Warsaw Medical University, 63A Zwirki i Wigury Street, 02-091 Warsaw, Poland

Postgraduate School of Molecular Medicine, Warsaw Medical University, 63A Zwirki i Wigury Street, 02-091 Warsaw, Poland
e-mail: aneta.manda-handzlik@wum.edu.pl

W. Bystrzycka and S. Sieczkowska
Student Scientific Group at the Department of Laboratory Diagnostics and Clinical Immunology of Developmental Age, Warsaw Medical University, 63A Zwirki i Wigury Street, 02-091 Warsaw, Poland

U. Demkow and O. Ciepiela
Department of Laboratory Diagnostics and Clinical Immunology of Developmental Age, Warsaw Medical University, 63A Zwirki i Wigury Street, 02-091 Warsaw, Poland

et al. 2009). It has been recently shown that antibiotics can also modulate the formation of neutrophil extracellular traps by bovine granulocytes (Jerjomiceva et al. 2014).

Neutrophil extracellular traps (NETs) have been described several years ago (Brinkmann et al. 2004) as a novel strategy used by neutrophils to fight and control infections. NETs are web-like structures composed of DNA and antimicrobial proteins, which entrap pathogens and create space with a high local concentration of antimicrobial factors. NETs are an efficient weapon released in response to wide range of microorganisms – bacteria, viruses, parasites and fungi. They are of critical role especially when pathogens are too large to be phagocytized (parasites, hyphae form of fungi), as granulocytes can size up pathogens and selectively release DNA and antimicrobial content to the environment (Branzk et al. 2014). However, uncontrolled neutrophil extracellular traps release entails some risk for the host. NETs content is highly immunogenic, normally stored within the cells, and when presented to the cells of immune system, can exacerbate autoimmune diseases, such as lupus, rheumatoid arthritis, or small vessel vasculitis (Pieterse et al. 2015; Pratesi et al. 2014; Sangaletti et al. 2012). Thus, it is crucial to keep the strict balance between NETs formation and clearance.

The aim of the present study was to investigate whether cefotaxime and gentamicin, which are among antibiotics known to affect phagocyte function, can influence NETs formation in humans.

2 Methods

2.1 Neutrophil Isolation

The study was approved by the Ethics Committee of the Medical University of Warsaw in Poland. Human blood was collected from healthy individuals. Informed, written consent was obtained from all subjects in the study.

Neutrophils were isolated using a density centrifugation method (Skrzeczynska-Moncznik et al. 2013). Rich-platelet plasma was discarded and replaced with 0.9 % solution of NaCl. Diluted buffy coat was then layered onto Histopaque 1077. After centrifugation, top layers (mononuclear cells and isolation media Histopaque 1077 layers) were carefully removed and the layer containing polymorphonuclear cells and red blood cell pellet was mixed with a 1 % solution of polyvinyl alcohol. Subsequently, after a 20-min incubation at room temperature, the upper phase containing alcohol and neutrophils was collected and centrifuged. The remaining red blood cells were lysed by hypotonic lysis and granulocytes were washed twice. The supernatant was removed and the cell pellet was suspended in a cell culture medium (RPMI 1640) without phenol red, supplemented with 10 mM HEPES (Gibco, Waltham, MA)).

2.2 NETs Quantification

Freshly isolated neutrophils were seeded into 96-well black plates (10^5 cells/well) and incubated for 2 h in 5 % CO_2 atmosphere at 37 °C with 1 μl/ml gentamicin, 10 μg/ml cefotaxime (G 1379, Sigma Aldrich, Saint Louis, USA), or the Roswell Park Memorial Institute (RPMI) 1640 Medium alone. Then, NETs formation was stimulated with phorbol myristate acetate (PMA, 100 nM) or calcium ionophore (CI, 4 μM), and the cells were incubated for 3 h in 5 % CO_2 atmosphere at 37 °C. Unstimulated neutrophils were used as controls. After incubation, 500mIU/ml of micrococcal nuclease (ThermoFisher Scientific, Waltham, MA) was added to detach DNA from the cell surface, and the reaction was stopped with 5 mM EDTA after 20 min incubation at 37 °C. The cells were centrifuged, the supernatant was collected and DNA release was measured in the presence of 100 nM Sytox green fluorescent dye (Life Technologies, Waltham, MA) in a FLUOstar OMEGA plate reader (BMG Labtech, Ortenberg, Germany).

2.3 NETs Visualization

Neutrophils were incubated in eight-well Lab-Tek chamber slides (2.5×10^4 cells/well; Nunc, Waltham, MA) with antibiotics and then NETs formation was stimulated as described above. Three hours post-stimulation samples were fixed with 4 % paraformaldehyde and DNA was stained with Sytox green (Life Technologies; Waltham, MA). NETs were visualized using a Leica DMi8 fluorescent microscope equipped with 40x magnification objective.

2.4 Statistical Analysis

Results are expressed as means ± SE of six independent experiments performed in triplicates. Multiple comparisons were made by one-way ANOVA followed by Dunn's *post-hoc* test. A p-value <0.05 was used to define statistical significance of differences. Statistical calculations were performed using a commercial package of GraphPad Prism v. 5.0 (GraphPad Software, La Jolla, CA).

3 Results

We isolated neutrophils from fresh blood of healthy volunteers by density gradient centrifugation. The cells were then treated with 1 μl/ml gentamicin or 10 μg/ml cefotaxime, and their influence on NETs formation was analyzed. As expected from other studies, both PMA and CI enhanced NETs formation compared with the unstimulated neutrophils. The enhancement was maniefested as an increase in DNA fluorescence and decondensation of nuclei (decondensation of chromatin is a step preceding DNA release to the extracellular space) accompanied by the formation of DNA-positive web-like threads (Fig. 1b, c, and f).

A quantitative measurement of DNA release by fluorometry demonstrates that from the two antibiotics tested, only gentamicin affected NETs release. The mean fluorescence in the PMA-stimulated neutrophils was 12875 ± 1257 RFU (relative fluorescence units) and in the presence of gentamicin it decreased to 9064 ± 1585 RFU ($p < 0.05$). Likewise, DNA release was diminished when NETs release was CI-stimulated. The mean fluorescence in the CI-treated samples was 6339 ± 1724 RFU and it decreased to 4040 ± 1351 RFU after inhibition with gentamicin ($p < 0.01$). Treatment with gentamicin (Fig. 1a, d, and g) diminished NETs formation when compared with that in the presence of the PMA and CI stimuli alone (Fig. 1a, c, and f) by 30.6 ± 8.6 % and 33.7 ± 19.6 %, respectively (Fig. 1a, c, d, f, and g). Cefotaxime did not influence NETs formation (Fig. 1a, e, and h). These observations were confirmed by fluorescent microscopy. We found that the number of decondensed nuclei and NETs was lower in the gentamicin-treated than untreated neutrophils.

4 Discussion

This study demonstrates that antibiotics may affect NETs release by human neutrophils. Neutrophil extracellular traps are among major mechanisms used by neutrophils to eradicate bacteria. They have been described primarily as a beneficial tool to entrap and kill pathogens extracellularly (Brinkmann et al. 2004) in the process called NETosis. Recently, intense research has addressed the role of NETs in health and disease. NETs release can contribute to the pathogenesis of several diseases, including autoimmune diseases such as systemic lupus erythematosus (Pieterse et al. 2015; Zawrotniak et al. 2015). Thus, a considerable part of the research is being directed toward identification of the agents capable of modulating NETs release. However, the interaction between antibiotics and release of NETs in humans has not yet been elucidated.

In the present study we tested two antibiotics, which belong to different groups: cefotaxime, a third-generation cephalosporin and gentamicin, an aminoglycoside. Both are used to treat many

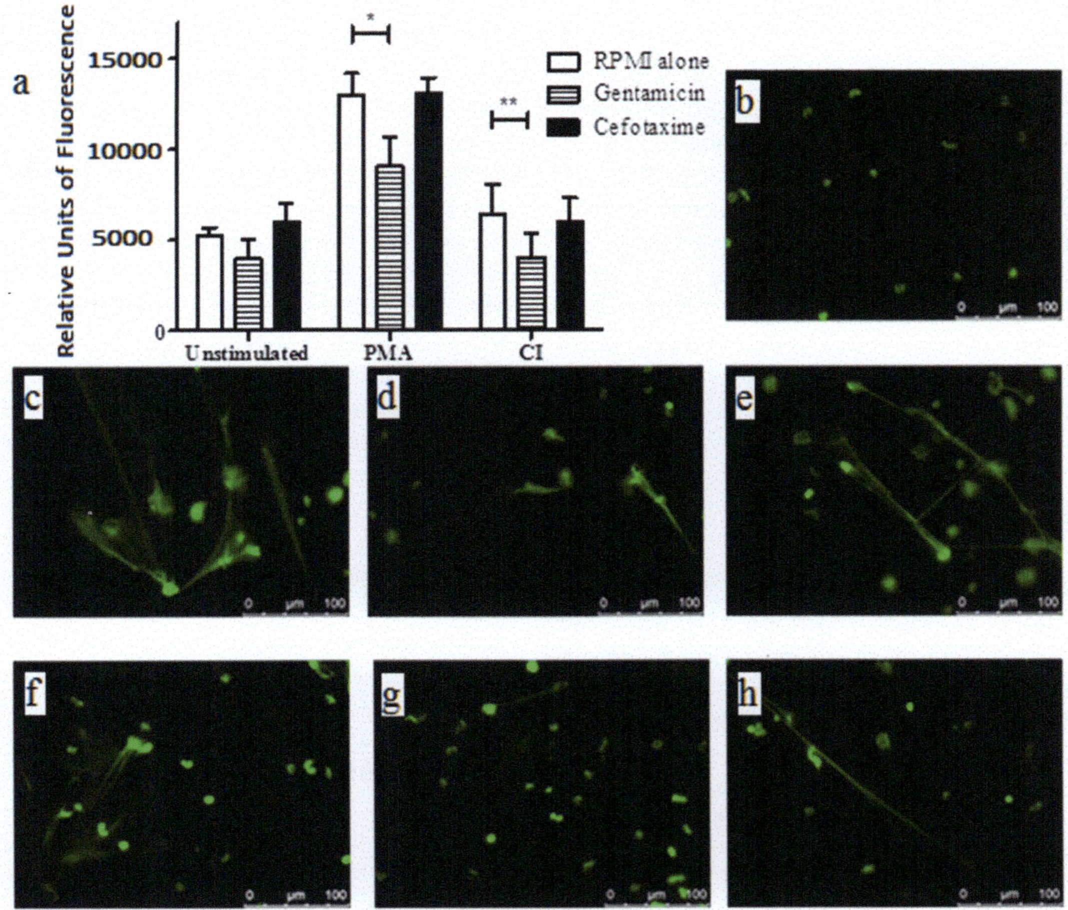

Fig. 1 (**a**) Effects of gentamicin and cefotaxime on NETs release from unstimulated (RPMI), and phorbol myristate acetate (PMA) and calcium ionophore (CI)-stimulated human neutrophils. DNA release was measured fluorometrically in the presence of extracellular DNA-binding dye Sytox green. The results represent 7 independent experiments performed in triplicates (see Methods for details). (**b**-**h**) Fluorescence microscopy of NETs: (**b**) unstimulated neutrophils, (**c**) PMA-stimulated neutrophils, (**d**) gentamicin-pretreated PMA-stimulated neutrophils, (**e**) cefotaxime-pretreated PMA-stimulated neutrophils, (**f**) CI-stimulated neutrophils, (**g**) gentamicin-pretreated CI-stimulated neutrophils, (**h**) cefotaxime-pretreated CI-stimulated neutrophils. Gentamicin decreased whereas cefotaxime did not influence NETs release from neutrophils. The images represent three independent experiments using blood from different donors;$^{*}p < 0.05$, $^{**}p < 0.01$

types of infections, including lung, urinary tract, or bone infections. Bacteria against which these antibiotics show activity include *Escherichia coli, Staphylococcus aureus, Streptococcus pneumoniae* (gentamicin and cefotaxime) and *Pseudomonas aeruginosa* (gentamicin), These bacteria can stimulate NETs release by granulocytes (Halverson et al. 2015; Mori et al. 2012). An interesting question arises of whether the antibiotics used to treat infections caused by bacteria triggering NETs release, apart from exerting a direct antimicrobial effect, could also modulate the antimicrobial strategy employed by specialized phagocyting cells.

Under the experimental conditions of the present study, gentamicin affected NETs release, while cefotaxime did not. We observed a diminished formation of NETs after gentamicin treatment in the stimulated neutrophils, regardless of the kind of stimulus used, be it CI or PMA. We demonstrate that the effect of antibiotics on NETs release varies depending on the kind of antibiotic. A similar study has been performed by Jerjomiceva et al. (2014) who tested the influence of the fluoroquinolone enrofloxacin on bovine granulocytes. Enrofloxacin is used for lactating cattle for treatment of mastitis induced by *Staphylococcus aureus* and *Escherichia coli*. The authors show that enrofloxacin induces the formation of bovine NETs and inhibits phagocytosis of *Staphylococcus aureus*. In the present study we did not notice similar effects as treatment with either antibiotic used failed to induce NETs release in resting neutrophils (Fig. 1a). The discrepancy between the results of the two studies is explicable by species differences and by structural differences in the antibiotics. Jerjomiceva et al. (2014) have found that enrofloxacin-induced NETs release is mediated by reactive oxygen species (ROS) and citrullination of histones; the two mechanisms that play a pivotal role in NETs formation (Keshari et al. 2013; Wang et al. 2009). Gentamicin is capable of increasing ROS formation (Martinez-Salgado et al. 2002). However, a ROS-linked mechanism can hardly be involved in the interaction between neutrophils releasing NETs and gentamicin in the present study. If it were we would have expected an increase, rather than decrease, in NETs release. Moreover, the effect of gentamicin on NETs release cannot be related to ROS formation as gentamicin affected NETosis triggered by both ROS-dependent (PMA) and ROS-independent (CI) stimuli.

Schilcher et al. (2014) have shown that clindamycin, a linkozamide antibiotic inhibiting protein synthesis, decreases the ability of *Staphylococcus aureus* to degrade neutrophil extracellular traps. A reduced NETs degradation is associated with inhibited nuclease activity. That shows that antibiotics can affect not only the ability of phagocytic cells to release NETs, but also modify bacteria defense systems against human innate immunity.

NETs are beneficial as they limit the area of inflammation and kill bacteria, but, on the other hand, the content of NETs can be presented to cells of the immune system and exacerbate autoimmune diseases like systemic lupus erythematosus. Our current results raise a nagging question of the clinical meaning of the antibiotic influence on NETs release. It can be suggested that it would be advisable in some patients to choose, if only feasible, antibiotics that are capable of increasing NETs formation and thus enhance the antimicrobial response. In others, suffering from NETs-related diseases, it would be optimal to use antibiotics that diminish NETosis.

In synopsis, we believe we have demonstrated that antibiotics influence NETs formation in humans. The finding is of clear therapeutic importance in infectious diseases in patients suffering from NET-related diseases.

Acknowledgments This study was supported by a Grant for Young Scientist no. 1WW/PM11/15/15 funded by the Dean of the First Faculty of Medicine, Warsaw Medical University in Warsaw, Poland.

Conflicts of Interest The authors declare no conflicts of interest in relation to this article.

References

Branzk N, Lubojemska A, Hardison SE, Wang Q, Gutierrez MG, Brown GD, Papayannopoulos V (2014) Neutrophils sense microbe size and selectively release neutrophil extracellular traps in response to large pathogens. Nat Immunol 15:1017–1025

Brinkmann V, Reichard U, Goosmann C, Fauler B, Uhlemann Y, Weiss DS, Weinrauch Y, Zychlinsky A (2004) Neutrophil extracellular traps kill bacteria. Science 303:1532–1535

Halverson TW, Wilton M, Poon KK, Petri B, Lewenza S (2015) DNA is an antimicrobial component of neutrophil extracellular traps. PLoS Pathog 11:e1004593

Jerjomiceva N, Seri H, Völlger L, Wang Y, Zeitouni N, Naim HY, von Köckritz-Blickwede M (2014) Enrofloxacin enhances the formation of neutrophil extracellular traps in bovine granulocytes. J Innate Immun 6:706–712

Keshari RS, Verma A, Barthwal MK, Dikshit M (2013) Reactive oxygen species-induced activation of ERK and p38 MAPK mediates PMA-induced NETs release from human neutrophils. J Cell Biochem 114:532–540

Martinez-Salgado C, Eleno N, Tavares P, Rodriguez-Barbero A, Garcia-Criado J, Bolaños JP, López-Novoa JM (2002) Involvement of reactive oxygen species on gentamicin-induced mesangial cell activation. Kidney Int 62:1682–1692

Minic S, Bojic M, Vukadinov J, Canak G, Fabri M, Bojic I (2009) Immunomodulatory actions of antibiotics. Med Pregl 62:327–330. (Article in Serbian)

Mori Y, Yamaguchi M, Terao Y, Hamada S, Ooshima T, Kawabata S (2012) Alpha-Enolase of Streptococcus pneumoniae induces formation of neutrophil extracellular traps. J Biol Chem 287:10472–10481

Pieterse E, Hofstra J, Berden J, Herrmann M, Dieker J, van der Vlag J (2015) Acetylated histones contribute to the immunostimulatory potential of neutrophil extracellular traps in systemic lupus erythematosus. Clin Exp Immunol 179:68–74

Pratesi F, Dioni I, Tommasi C, Alcaro MC, Paolini I, Barbetti F, Boscaro F, Panza F, Puxeddu I, Rovero P, Migliorini P (2014) Antibodies from patients with rheumatoid arthritis target citrullinated histone 4 contained in neutrophils extracellular traps. Ann Rheum Dis 73:1414–1422

Roszkowski W, Ko HL, Roszkowski K, Jeljaszewicz J, Pulverer G (1985) Antibiotics and immunomodulation: effects of cefotaxime, amikacin, mezlocillin, piperacillin and clindamycin. Med Microbiol Immunol 173:279–289

Sangaletti S, Tripodo C, Chiodoni C, Guarnotta C, Cappetti B, Casalini P, Piconese S, Parenza M, Guiducci C, Vitali C, Colombo MP (2012) Neutrophil extracellular traps mediate transfer of cytoplasmic neutrophil antigens to myeloid dendritic cells toward ANCA induction and associated autoimmunity. Blood 120:3007–3018

Schilcher K, Andreoni F, Uchiyama S, Ogawa T, Schuepbach RA, Zinkernagel AS (2014) Increased neutrophil extracellular trap-mediated Staphylococcus aureus clearance through inhibition of nuclease activity by clindamycin and immunoglobulin. J Infect Dis 210:473–482

Skrzeczynska-Moncznik J, Wlodarczyk A, Banas M, Kwitniewski M, Zabieglo K, Kapinska-Mrowiecka M, Dubin A, Cichy J (2013) DNA structures decorated with cathepsin G/secretory leukocyte proteinase inhibitor stimulate IFNI production by plasmacytoid dendritic cells. Am J Clin Exp Immunol 2:186–194

Tentori L, Graziani G, Porcelli SA, Sugita M, Brenner MB, Madaio R, Bonmassar E, Giuliani A, Aquino A (1998) Rifampin increases cytokine-induced expression of the CD1b molecule in human peripheral blood monocytes. Antimicrob Agents Chemother 42:550–554

Wang Y, Li M, Stadler S, Correll S, Li P, Wang D, Hayama R, Leonelli L, Han H, Grigoryev SA, Allis CD, Coonrod SA (2009) Histone hypercitrullination mediates chromatin decondensation and neutrophil extracellular trap formation. J Cell Biol 184:205–213

Zawrotniak M, Kozik A, Rapala-Kozik M (2015) Selected mucolytic, anti-inflammatory and cardiovascular drugs change the ability of neutrophils to form extracellular traps (NETs). Acta Biochim Pol 62:465–473

Ziegeler S, Raddatz A, Hoff G, Buchinger H, Bauer I, Stockhausen A, Sasse H, Sandmann I, Horsch S, Rensing H (2006) Antibiotics modulate the stimulated cytokine response to endotoxin in a human ex vivo, *in vitro* model. Acta Anaesthesiol Scand 50:1103–1110

Advs Exp. Medicine, Biology - Neuroscience and Respiration (2017) 27: 53–62
DOI 10.1007/5584_2016_42

Published online: 13 July 2016

Sensitivity of Next-Generation Sequencing Metagenomic Analysis for Detection of RNA and DNA Viruses in Cerebrospinal Fluid: The Confounding Effect of Background Contamination

Iwona Bukowska-Ośko, Karol Perlejewski, Shota Nakamura, Daisuke Motooka, Tomasz Stokowy, Joanna Kosińska, Marta Popiel, Rafał Płoski, Andrzej Horban, Dariusz Lipowski, Kamila Caraballo Cortés, Agnieszka Pawełczyk, Urszula Demkow, Adam Stępień, Marek Radkowski, and Tomasz Laskus

Abstract

Next-generation sequencing (NGS) followed by metagenomic analysis enables the detection and identification of known as well as novel pathogens. It could be potentially useful in the diagnosis of encephalitis, caused by a variety of microorganisms. The aim of the present study was to evaluate the sensitivity of isothermal RNA amplification (Ribo-SPIA) followed by NGS metagenomic analysis in the detection of human immunodeficiency virus (HIV) and herpes simplex virus (HSV) in cerebrospinal

The original version of this chapter has been revised. An erratum to this chapter can be found at DOI 10.1007/978-3-319-44488-8_11

I. Bukowska-Ośko, K. Perlejewski (✉), M. Popiel, K. Caraballo Cortés, A. Pawełczyk, M. Radkowski, and T. Laskus
Department of Immunopathology of Infectious and Parasitic Diseases, Warsaw Medical University, 3C Pawińskiego St, 02-106 Warsaw, Poland
e-mail: karol.perlejewski@wum.edu.pl

S. Nakamura and D. Motooka
Department of Infection Metagenomics, Genome Information Research Center, Research Institute for Microbial Diseases, Osaka University, 3-1 Yamadaoka, Suita, Osaka, Japan

T. Stokowy
Department of Clinical Science, Bergen University, 5021 Bergen, Norway

J. Kosińska and R. Płoski
Department of the Medical Genetics, Warsaw Medical University, 3C Pawińskiego St, 02-106 Warsaw, Poland

A. Horban and D. Lipowski
Municipal Hospital for Infectious Diseases, 37 Wolska St, 01-201 Warsaw, Poland

U. Demkow
Department of Laboratory Medicine and Clinical Immunology of Developmental Age, Warsaw Medical University, 24 Marszałkowska St, 00-576 Warsaw, Poland

A. Stępień
Department of Neurology, Military Institute of Medicine, 128 Szaserów St, 04-141 Warsaw, Poland

fluid (CSF). Moreover, we analyzed the contamination background. We detected 10^2 HIV copies and 10^3 HSV copies. The analysis of control samples (two water samples and one CSF sample from an uninfected patient) revealed the presence of human DNA in the CSF sample (91 % of all reads), while the dominating sequences in water were qualified as 'other', related to plants, plant viruses, and synthetic constructs, and constituted 31 % and 60 % of all reads. Bacterial sequences represented 5.9 % and 21.4 % of all reads in water samples and 2.3 % in the control CSF sample. The bacterial sequences corresponded mainly to *Psychrobacter, Acinetobacter,* and *Corynebacterium* genera. In conclusion, Ribo-SPIA amplification followed by NGS metagenomic analysis is sensitive for detection of RNA and DNA viruses. Contamination seems common and thus the results should be confirmed by other independent methods such as RT-PCR and PCR. Despite these reservations, NGS seems to be a promising method for the diagnosis of viral infections.

Keywords

Bacteria • Cerebrospinal fluid • DNA • Encephalitis • Next-generation sequencing • Metagenomics analysis • Pathogens • Viruses

1 Introduction

Next-generation sequencing (NGS) is frequently used to analyze microbial evolution, pathogen transmission patterns, and to identify drug-resistant variants (Frey et al. 2014; Barzon et al. 2013; Capobianchi et al. 2013; Dunne et al. 2012). NGS is also employed to identify etiologic agents in various infections (Miller et al. 2013; Yozwiak et al. 2012; Virgin and Todd 2011; Nakamura et al. 2009). Metagenomic sequencing provides insight into microbial composition of a sample, enables the detection of non-culturable and highly variable organisms, as well as emerging and novel pathogens (Barzon et al. 2013; Capobianchi et al. 2013; Frey et al. 2014).

Infections of the central nervous system are caused by a variety of microorganisms including viruses, bacteria, fungi, and parasites. Until now, more than 100 different agents have been identified as capable of causing encephalitis (Granerod et al. 2010a; Glaser et al. 2003). The most common viral agents causing encephalitis in Europe are herpes simplex virus (HSV), arboviruses (tick-borne encephalitis virus), and enteroviruses (Paradowska-Stankiewicz and Piotrowska 2014; Granerod and Crowcroft 2007; Glaser et al. 2006). However, etiology remains unknown in 40–60 % of cases, which could be due to a wide range of factors such as suboptimal time of sample collection, large number of potential causative microorganisms, or the existence of yet unknown pathogens (Silva 2013; Granerod et al. 2010b). Routine diagnostics, which typically includes serological and molecular tests, enables the detection only of the most common pathogens (Rasool et al. 2014; Solomon et al. 2007; Steiner et al. 2005; Chaudhuri and Kennedy 2002).

NGS could be a valuable tool in the diagnosis of encephalitis as it could detect causative pathogens even when the routine tests fail (Wilson et al. 2014; Tan le et al. 2013). In the current study we analyzed the sensitivity of NGS, preceded by isothermal RNA amplification (Ribo-SPIA), for the detection of HIV and HSV in the cerebrospinal fluid (CSF).

2 Methods

2.1 Patients and Samples

The study was approved by the Internal Review Board of Warsaw Medical University in Poland. Cerebrospinal fluid (CSF) was obtained from a HIV-positive 28-year-old woman hospitalized with sepsis and impaired consciousness at Warsaw Municipal Hospital for Infectious Diseases, in March 2012. HIV viral load in serum and CSF was 10^6 copies and 1.5×10^5 copies/ml, respectively (ABBOTT Real Time PCR HIV CE Abbott Molecular Inc., Des Plaines, IL). A control CSF sample was obtained as part of a routine diagnostic procedure from a 69-year-old man with motor neuron disease.

HSV-1 positive samples consisted of DNA isolated from Vero cell line (ATCC CCL81) and provided by the Reference Virus Isolation Service Laboratory (New York, NY). Viral load was 4.4×10^3 copy/μl (LightCycler HSV 1/2 Qualitative Kit Roche Diagnostics, Mannheim, Germany). As a negative control (sample W1 and W2) ultrapure water (Water Molecular Biology Reagent, Sigma Aldrich) was used.

The HIV-positive CSF sample was diluted in the HIV-negative CSF to obtain concentrations of 10^4, 10^3, 10^2, and 10^1 viral copies per reaction. Serial dilutions of HSV were obtained by spiking negative CSF with control HSV DNA. Likewise, dilutions were adjusted to 10^4, 10^3, 10^2, and 10^1 viral copies per reaction.

2.2 Nucleic Acid Isolation and Amplification

Total RNA was extracted from 250 μl of CSF by the Chomczynski method (Chomczynski and Sacchi 1987). The yield and purity of isolated RNA was estimated using a NanoDrop Spectrophotometer and Qubit 2.0 Fluorometer with RNA HS Assay Kit (Thermo Scientific, Waltham, MA).

The extracted RNA and DNA were subjected to isothermal RNA amplification (Ribo-SPIA) using commercially available kit RNA-Seq Ovation V2 kit (NuGEN, San Carlos, CA). This system requires a minimal amount of input RNA (500 pg) and generates a smaller background of human sequences then other commercially available kits (Malboeuf et al. 2013; Kurn et al. 2005). The manufacturer's protocol was closely followed and the only modification was a substitution of random primers for oligo-dT mix. The amplified products were purified using AMPure XP beads (Beckman Coulter, Pasadena, CA) and measured on a Qubit 2.0 Fluorometer (Life Technology, Carlsbad, CA). In our earlier study we showed that the above approach enables the detection of both RNA and DNA pathogens (Perlejewski et al. 2015).

2.3 NGS Library Construction, Sequencing and Bioinformatic Analysis

The library for each sample was generated using an Nextera XT Kit (Illumina, San Diego, CA) according to the manufacturer's protocol. Briefly, cDNA was fragmented using transposon-based method followed by dual indexes addition with PCR. To eliminate short cDNA inserts within sequences and to increase the final length of sequences in the library, the following modifications were introduced: (1) final concentration of DNA digesting enzyme (transposase) was reduced by one fourth (2) PCR products were purified twice using 1.8 and 1.6 volumes of AMPure XP beads.

The quality and average length of sequence library for each sample was assessed using a Bioanalyzer (Agilent Technologies, Santa Clara, CA) with a high sensitivity DNA Assay. Six indexed samples were pooled equimolarly and sequenced on a single lane of the Illumina HiSeq 1500 (100 base paired-end reads).

A bioinformatic analysis was conducted as described elsewhere (Perlejewski et al. 2015). In short, raw Illumina high-throughput sequencing reads were analyzed with a pipeline that included quality control, trimming, reference alignment to the human reference sequence hg19, and the whole unfiltered NCBI-nt database and were followed by aligned sequence counting and annotation.

3 Results

3.1 Sensitivity of RNA and DNA Virus Detection

Sequencing of four samples containing different HSV-1 loads (10^1, 10^2, 10^3, and 10^4 viral copies per reaction) provided 135,846,381 (29.4–37.3 million per sample) reads. The average number of reads per sample was 33,961,595. Similar numbers of reads were obtained for HIV samples 130,7834,29 (25.5–38.8 million per sample) and the average number of sequences was 32,695,857. After trimming there were 130,232,947 (28.2–32.2 million per sample) reads for samples containing HSV-1 and 128,741,208 (25.1–38.0 million per sample) reads for HIV-positive samples.

Viral sequences represented 0.001–0.620 % of all reads in HIV-positive samples and 0.040–2.540 % of all reads in HSV-positive samples. The results of sequence alignments are shown in Table 1 and Fig. 1.

In the HIV RNA-positive CSF samples, the highest number of HIV sequences were detected in a sample containing 10^4 viral copies. In this sample, 96.2 % of all viral reads were positively aligned to the HIV genome (0.6 % of all reads, coverage 97.0 % of HIV genome). In the samples containing 10^3 and 10^2 viral copies, 88.5 % and 73.8 % of viral sequences, respectively, aligned to the HIV genome constituted 0.022 % and 0.006 % of all reads (coverage 90.2 % and 30.0 % of HIV genome, respectively). HIV sequences were not detected in the sample with the lowest viral copy number (10^1).

The presence of HSV-1 DNA was detected only in the two samples with the highest viral titer of 10^3 and 10^4 and constituted 0.13 % and 25.90 % of all viral reads and 0.003 % and 0.028 % of all reads, respectively. The HSV-specific reads covered 3.6 % and 13.0 % of the viral genome at the concentrations of 10^3 and 10^4 copies/reaction, respectively. HSV-1 sequences were not detected in CSF containing 10^1 and 10^2 viral copies.

3.2 Microbial Contamination

The most abundant in both HIV and HSV-positive samples were human sequences (91.46–97.35 % of all sequences) followed by bacterial sequences (1.04–4.14 % and 0.91–3.69 %); (Table 1). The results of metagenomic analysis of two sterile water samples (W1 and W2) and one control CSF sample (N) are shown in Tables 2 and 3. The analysis of the three negative controls was carried out in three independent sequencing runs.

After trimming, there were 1,521,154 reads for sample W1, 7,192,898 reads for sample W2, and 40,340,940 reads for sample N (negative CSF). Human DNA was common in the CSF control sample (90.813 % of all reads), whereas the sequences identified as ‘other’ (sequences related to plants, plant viruses, and synthetic constructs) dominated in both water samples (31.419 % and 60.326 % of all reads). Human sequences were less abundant in water (11.0 % and 13.0 %) and the ‘other’ sequences constituted only 3.116 % of all reads in sample N.

Bacterial sequences represented 5.85 % and 21.40 % of all reads in water samples and 2.3o% in the control sample N. The most frequent bacterial sequences corresponded to *Psychrobacter* (sample W1), *Acinobacter* (sample W2), and *Corynebacterium* (sample N). Other common species included genera *Pseudomonas*, *Streptococcus, Staphylococcus, Escherichia, Corynebacterium, Bacillus, Propionibacterium* and *Micrococcus* (Table 3). A total of 28 different bacterial genera were common to all analyzed samples. Viral, fungal, and protozoan reads were less numerous and did not exceed 1 % in any sample.

Table 1 Bioinformatic analysis of next-generation sequencing (NGS) data from samples containing serial dilutions of herpes simplex virus (HSV-1) and human immunodeficiency virus (HIV)

Virus	HIV				HSV-1			
Viral copy number (per reaction)	10^1	10^2	10^3	10^4	10^1	10^2	10^3	10^4
All reads	**28,061,486**	**25,451,203**	**38,824,869**	**38,445,871**	**29,427,549**	**35,563,729**	**37,265,053**	**33,590,050**
Reads after trimming	27,521,179	25,130,211	38,030,780	38,059,038	28,167,940	34,159,264	35,710,638	32,195,105
Human	26,790,866	23,491,647	37,058,865	36,195,659	26,490,780	33,227,773	32,660,536	31,208,000
	(97.35 %)	(93.48 %)	(97.44 %)	(95.10 %)	(94.05 %)	(97.27 %)	(91.46 %)	(96.93 %)
Viral	2452	2035	9860	235,243	113,081	14,679	908,566	34,657
	(0.01 %)	(0.01 %)	(0.03 %)	(0.62 %)	(0.40 %)	(0.04 %)	(2.54 %)	(0.11)
Bacterial	285,519	1,039,411	409,000	574,836	1,040,489	552,050	1,123,961	294,172
	(1.04 %)	(4.14 %)	(1.08 %)	(1.51 %)	(3.69 %)	(1.62 %)	(3.15 %)	(0.91 %)
Fungal	41,242	111,571	92,310	118,438	48,506	26,304	141,358	52,367
	(0.15 %)	(0.26 %)	(0.24 %)	(0.31 %)	(0.17 %)	(0.08 %)	(0.40 %)	(0.16 %)
Archeal	155	13	55	501	35	18	192	52
	(0.00 %)	(0.00 %)	(0.00 %)	(0.00 %)	(0.00 %)	(0.00 %)	(0.00 %)	(0.00 %)
Protozoan	27,373	38,223	37,608	60,627	22,476	14,057	102,683	94,557
	(0.10 %)	(0.15 %)	(0.10 %)	(0.16 %)	(0.08 %)	(0.04 %)	(0.29 %)	(0.29 %)
Other[a]	155,230	188,205	183,186	385,748	194,950	143,329	453,406	270,508
	(0.56 %)	(0.75 %)	(0.48 %)	(1.01 %)	(0.69 %)	(0.42 %)	(1.27 %)	(0.84 %)
No match	218,342	259,106	239,896	487,986	257,623	181,054	319,936	240,792
	(0.79 %)	(1.03 %)	(0.63 %)	(1.28 %)	(0.91 %)	(0.53 %)	(0.90 %)	(0.75 %)
Identified viruses (number of reads)	0	HIV (1501)	HIV (8731)	HIV (226,290)	0	0	HSV-1 (1151)	HSV-1 (8989)
		0.006 %[b]	0.022 %[b]	0.590 %[b]			0.003 %[b]	0.028 %[b]
Viral genome coverage	0	30.0 %	90.2 %	97.0 %	0	0	3.6 %	13.0 %

Sequences were compared to the NCBI-nt database
[a]Sequences related to plants, plant viruses, and synthetic constructs
[b]Percentage of all reads after trimming

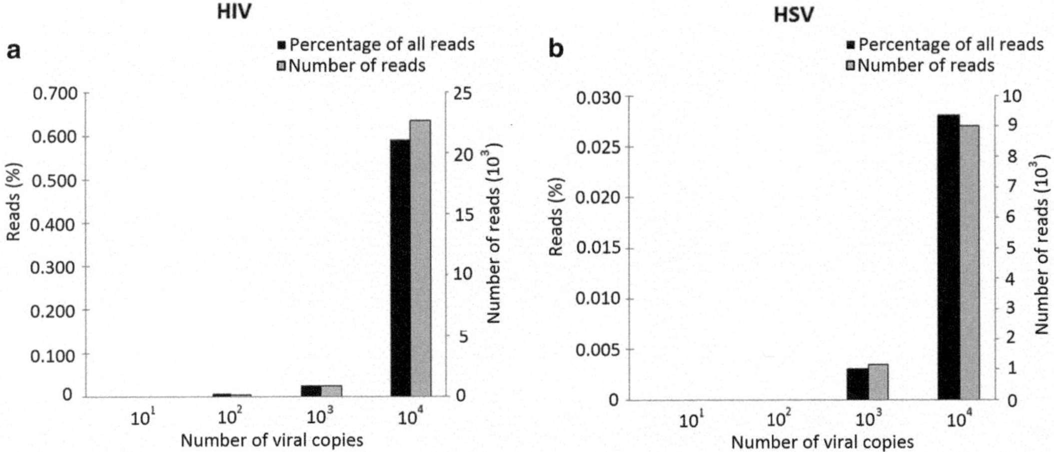

Fig. 1 Metagenomic analysis of next-generation sequencing (NGS) of samples containing different human immunodeficiency virus (HIV) and herpes simplex virus (HSV-1) loads (10^1, 10^2, 10^3, and 10^4 viral copies per reaction). Percentages reflect the proportion of HIV and HSV-1 reads in regard to all reads after trimming

Table 2 Next-generation sequencing (NGS) of negative controls: water (W1 and W2) and cerebrospinal fluid (CSF) samples from a patient with motor neuron disease (MND)

Sample	W1	W2	MND
Reads after trimming (n)	1,521,154	7,192,898	40,340,940
Human	168,102 (11.051 %)	967,092 (13.445 %)	36,634,692 (90.813 %)
Viral	711 (0.047 %)	36,231 (0.500 %)	6878 (0.017 %)
Bacterial	819,135 (5.850 %)	1,540,184 (21.410 %)	922,862 (2.288 %)
Fungal	8366 (0.550 %)	165,873 (2.306 %)	258,635 (0.641 %)
Archeal	0	0	500 (0.001 %)
Protozoan	69 (0.005 %)	28,830 (0.401 %)	106,155 (0.263 %)
Other[a]	477,933 (31.419 %)	4,339,215 (60.326 %)	1,256,859 (3.116 %)
No match	46,838 (3.079 %)	115,473 (1.605 %)	1,154,359 (2.862 %)

Sequences were compared to the NCBI-nt database
[a]Sequences related to plants, plant viruses, and synthetic constructs

4 Discussion

Detection of individual pathogens in infections of the central nervous system is often impractical due to the sheer number of potential infectious agents. NGS represents a novel and promising tool in microbial diagnostics, as it can detect multiple pathogens concurrently. NGS techniques, combined with a metagenomics approach, have enabled the identification of new arenavirus (Palacios et al. 2008) and previously unknown species of *Bundibugyo ebolavirus* (Towner et al. 2008). Other examples of a successful use of NGS were the identification of *Leptospira* in CSF of a 14-year-old patient with encephalitis of unknown etiology (Wilson

Table 3 Most frequently identified species/genera in the cerebrospinal fluid (CSF) sample from a control patient with motor neuron disease (MND) and in two water samples (W1 and W2)

Sample	Viruses	Bacteria[a]	Fungi[a]	Protozoa[a]
MND	None	Corynebacterium (96,345)	Funneliformis (19,789)	Besnoitia (54,232)
		Pseudomonas (82,301)	Malassezia (14,005)	Nannochloropsis (2028)
		Staphylococcus (64,799)	Glomus (13,976)	Phytophthora (1647)
		Bacillus (61,580)	Cladosporium (10,127)	Plasmodium (1270)
		Escherichia (41,394)	Melampsora (5663)	Albugo (1197)
W1	Influenza C (670)	Psychrobacter (307,551)	Phialocephala subalpine (1202)	None
	HCV (36)	Staphylococcus (128,538)	Malassezia sympodialis (731)	
	Pepino Mosaic Virus (3)	Corynebacterium (75,297)	Leptosphaeria maculans (682)	
		Micrococcus (35,496)	Parastagonospora nodorum (680)	
		No taxonomic data (29,436)	Arthroderma obtusum (562)	
W2	None	Acinetobacter (215,184)	Malassezia restricta (5967)	Albugo laibachii (4645)
		Streptococcus (103,489)	Malassezia globose (5882)	Phytophthora megasperma (2835)
		Corynebacterium (95,912)	Penicillium rubens (5828)	Nannochloropsis oceanica (2786)
		Micrococcus (90,144)	Polycephalomyces formosus (4732)	Peronospora aquatic (2427)
		Staphylococcus (75,621)	Aspergillus fumigatus (4505)	Amphifilidae sp. H-1 (1736)

HSV-1 herpes simplex virus type 1, *HIV* human immunodeficiency virus, *HCV* hepatitis C virus
[a]Five most numerous genera

et al. 2014) and the identification of cyclovirus CyCV-VN as a novel etiological factor in encephalitis (Tan le et al. 2013).

In the present investigation, sensitivity of NGS, preceded by isothermal RNA amplification (Ribo-SPIA), was 10^3 copies for HSV-1 and 10^2 copies for HIV in CSF. A similar sensitivity (using Ribo-SPIA and Illumina platform HiSeq) with respect to RNA viruses has been reported by Malboeuf et al. (2013), who detected 10^2 copies of HIV, West Nile virus (WNV), and the respiratory syncytial virus (RSV) in plasma samples. A head-to-head comparison of sensitivity of different NGS protocols is rarely done, but the Illumina platforms show the highest efficiency and sequencing depth when compared to such other NGS platform as Ion PGM Torrent and pyrosequencing-based platforms (Frey et al. 2014; Lecuit and Eloit 2014; Cheval et al. 2011).

Nonetheless, short reads generated by Illumina platforms make the assembly of a full length of microbial genomes challenging (Quinones-Mateu et al. 2014; Virgin and Todd 2011). That is particularly important for the detection of unknown pathogens, where *de novo* assembly is necessary (Miyamoto et al. 2014). In addition, high coverage rate of a reconstructed genome may facilitate the interpretation of results. Currently, there are no established criteria for the detection of pathogens by NGS. Palacios et al. (2008) have identified astrovirus infection in a transplant patient after detection of 14 sequences that constituted only 0.0135 % of all reads, while in a study of Feng et al. (2008) the identification of Merkel cell polyomavirus has based on the detection of only 2 viral sequences, constituting just 0.0005 % of all reads. In the present investigation, we were able to detect the presence of HSV and HIV sequences, constituting 0.003 % and 0.006 %, of all reads, respectively.

Contamination is a common problem in the metagenomic analysis (Dickins et al. 2014; Lusk 2014; Strong et al. 2014; Malboeuf et al. 2013; Kircher et al. 2012; Clement-Ziza et al. 2009). It could be a cross-contamination between samples during the NGS library preparation, as well as due to an incorrect process of sample demultiplexing or formation of artificial products. In the current study, bacterial sequences and reads classified as 'other' (sequences related to plants, plant viruses, and synthetic constructs) were identified in all samples, but they were particularly common in water (Table 2). Generation of artifacts is observed after genome amplification, and is particularly common for samples with a low DNA and RNA load (Lusk 2014; Perlejewski et al. 2015). A low amount of RNA or DNA present in CSF increases the likelihood of an artifact appearance in the metagenomics analysis.

In accord with other studies we identified *Acinetobacter* and *Psychobacter* sequences as the most common bacterial reads. These bacterial genera have been recently reported as the dominant contamination in the metagenomics analysis. The source of the bacterial sequences could be reagents (Salter et al. 2014; Shen et al. 2006; Newsome et al. 2004), contamination inherent to sequencing platforms, and errors during bioinformatic analysis (Laurence et al. 2014; Barzon et al. 2013; Capobianchi et al. 2013). To reduce the impact of contamination and sequencing errors on the interpretation of results, it has been recommended to use negative controls, repeat analysis in independent experiments, and catalog batches of all reagents (Lusk 2014; Salter et al. 2014; Leek et al. 2010). A bioinformatic approach taking into account the most common bacterial contaminations and making a statistical assessment of the presence of causative agents could also be helpful to this end (Naccache et al. 2014; Strong et al. 2014).

In conclusion, using Ribo-SPIA amplification followed by NGS metagenomic analysis we were able to identify in the cerebrospinal fluid the presence of 10^2 copies of HIV and 10^3 copies of HSV per reaction. Contamination seems to be common in the metagenomic analysis and the results should be confirmed by an independent method such as RT-PCR or PCR. Despite these reservations, NGS seems a promising method in the diagnosis of viral infections.

Acknowledgments This study was supported by grants from the Foundation for Polish Science – POMOST/2013-7/2 and from the Polish National Science Center (NCN) – N/N401/646940.

Conflict of Interest The authors declare no conflicts of interest in relation to this article.

References

Barzon L, Lavezzo E, Costanzi G, Franchin E, Toppo S, Palu G (2013) Next-generation sequencing technologies in diagnostic virology. J Clin Virol 58 (2):346–350

Capobianchi MR, Giombini E, Rozera G (2013) Next-generation sequencing technology in clinical virology. Clin Microbiol Infect 19(1):15–22

Chaudhuri A, Kennedy PG (2002) Diagnosis and treatment of viral encephalitis. Postgrad Med J 78 (924):575–583

Cheval J, Sauvage V, Frangeul L, Dacheux L, Guigon G, Dumey N, Pariente K, Rousseaux C, Dorange F, Berthet N, Brisse S, Moszer I, Bourhy H, Manuguerra CJ, Lecuit M, Burguiere A, Caro V, Eloit M (2011) Evaluation of high-throughput sequencing for identifying known and unknown viruses in biological samples. J Clin Microbiol 49(9):3268–3275

Chomczynski P, Sacchi N (1987) Single-step method of RNA isolation by acid guanidinium thiocyanate-phenol-chloroform extraction. Anal Biochem 162 (1):156–159

Clement-Ziza M, Gentien D, Lyonnet S, Thiery JP, Besmond C, Decraene C (2009) Evaluation of methods for amplification of picogram amounts of total RNA for whole genome expression profiling. BMC Genomics 10:246

Dickins B, Rebolledo-Jaramillo B, Su MS, Paul IM, Blankenberg D, Stoler N, Makova KD, Nekrutenko A (2014) Controlling for contamination in re-sequencing studies with a reproducible web-based phylogenetic approach. Biotechniques 56(3):134–136, 138–141

Dunne WM, Westblade LF, Ford B (2012) Next-generation and whole-genome sequencing in the diagnostic clinical microbiology laboratory. Eur J Clin Microbiol Infect Dis 31(8):1719–1726

Feng H, Shuda M, Chang Y, Moore PS (2008) Clonal integration of a polyomavirus in human Merkel cell carcinoma. Science 319(5866):1096–1100

Frey KG, Herrera-Galeano JE, Redden CL, Luu TV, Servetas SL, Mateczun AJ, Mokashi VP, Bishop-Lilly KA (2014) Comparison of three next-generation sequencing platforms for metagenomic sequencing and identification of pathogens in blood. BMC Genomics 15:96

Glaser CA, Gilliam S, Schnurr D, Forghani B, Honarmand S, Khetsuriani N, Fischer M, Cossen CK, Anderson LJ, California Encephalitis (2003) In search of encephalitis etiologies: diagnostic challenges in the California Encephalitis Project, 1998–2000. Clin Infect Dis 36(6):731–742

Glaser CA, Honarmand S, Anderson LJ, Schnurr DP, Forghani B, Cossen CK, Schuster FL, Christie LJ, Tureen JH (2006) Beyond viruses: clinical profiles and etiologies associated with encephalitis. Clin Infect Dis 43(12):1565–1577

Granerod J, Crowcroft NS (2007) The epidemiology of acute encephalitis. Neuropsychol Rehabil 17 (4–5):406–428

Granerod J, Cunningham R, Zuckerman M, Mutton K, Davies NW, Walsh AL, Ward KN, Hilton DA, Ambrose HE, Clewley JP, Morgan D, Lunn MP, Solomon T, Brown DW, Crowcroft NS (2010a) Causality in acute encephalitis: defining aetiologies. Epidemiol Infect 138(6):783–800

Granerod J, Tam CC, Crowcroft NS, Davies NW, Borchert M, Thomas SL (2010b) Challenge of the unknown. A systematic review of acute encephalitis in non-outbreak situations. Neurology 75 (10):924–932

Kircher M, Sawyer S, Meyer M (2012) Double indexing overcomes inaccuracies in multiplex sequencing on the Illumina platform. Nucleic Acids Res 40(1):e3

Kurn N, Chen P, Heath JD, Kopf-Sill A, Stephens KM, Wang S (2005) Novel isothermal, linear nucleic acid amplification systems for highly multiplexed applications. Clin Chem 51(10):1973–1981

Laurence M, Hatzis C, Brash DE (2014) Common contaminants in next-generation sequencing that hinder discovery of low-abundance microbes. PLoS One 9(5):e97876

Lecuit M, Eloit M (2014) The diagnosis of infectious diseases by whole genome next generation sequencing: a new era is opening. Front Cell Infect Microbiol 4:25

Leek JT, Scharpf RB, Bravo HC, Simcha D, Langmead B, Johnson WE, Geman D, Baggerly K, Irizarry RA (2010) Tackling the widespread and critical impact of batch effects in high-throughput data. Nat Rev Genet 11(10):733–739

Lusk RW (2014) Diverse and widespread contamination evident in the unmapped depths of high throughput sequencing data. PLoS One 9(10):e110808

Malboeuf CM, Yang X, Charlebois P, Qu J, Berlin AM, Casali M, Pesko KN, Boutwell CL, DeVincenzo JP, Ebel GD, Allen TM, Zody MC, Henn MR, Levin JZ (2013) Complete viral RNA genome sequencing of ultra-low copy samples by sequence-independent amplification. Nucleic Acids Res 41(1):e13

Miller RR, Montoya V, Gardy JL, Patrick DM, Tang P (2013) Metagenomics for pathogen detection in public health. Genome Med 5(9):81

Miyamoto M, Motooka D, Gotoh K, Imai T, Yoshitake K, Goto N, Iida T, Yasunaga T, Horii T, Arakawa K, Kasahara M, Nakamura S (2014) Performance comparison of second- and third-generation sequencers using a bacterial genome with two chromosomes. BMC Genomics 15:699

Naccache SN, Federman S, Veeraraghavan N, Zaharia M, Lee D, Samayoa E, Bouquet J, Greninger AL, Luk KC, Enge B, Wadford DA, Messenger SL, Genrich GL, Pellegrino K, Grard G, Leroy E, Schneider BS, Fair JN, Martinez MA, Isa P, Crump JA, DeRisi JL, Sittler T, Hackett J, Miller S, Chiu CY (2014) A cloud-compatible bioinformatics pipeline for ultrarapid pathogen identification from next-generation sequencing of clinical samples. Genome Res 24(7):1180–1192

Nakamura S, Yang CS, Sakon N, Ueda M, Tougan T, Yamashita A, Goto N, Takahashi K, Yasunaga T, Ikuta K, Mizutani T, Okamoto Y, Tagami M, Morita R, Maeda N, Kawai J, Hayashizaki Y, Nagai Y, Horii T, Iida T, Nakaya T (2009) Direct metagenomic detection of viral pathogens in nasal and fecal specimens using an unbiased high-throughput sequencing approach. PLoS One 4(1), e4219

Newsome T, Li BJ, Zou N, Lo SC (2004) Presence of bacterial phage-like DNA sequences in commercial Taq DNA polymerase reagents. J Clin Microbiol 42 (5):2264–2267

Palacios G, Druce J, Du L, Tran T, Birch C, Briese T, Conlan S, Quan PL, Hui J, Marshall J, Simons JF, Egholm M, Paddock CD, Shieh WJ, Goldsmith CS, Zaki SR, Catton M, Lipkin WI (2008) A new arenavirus in a cluster of fatal transplant-associated diseases. N Engl J Med 358(10):991–998

Paradowska-Stankiewicz I, Piotrowska A (2014) Meningitis and encephalitis in Poland in 2012. Przegl Epidemiol 68(2):21–218, 333–216

Perlejewski K, Popiel M, Laskus T, Nakamura S, Motooka D, Stokowy T, Lipowski D, Pollak A, Lechowicz U, Caraballo Cortes K, Stepien A, Radkowski M, Bukowska-Osko I (2015) Next-generation sequencing (NGS) in the identification of encephalitis-causing viruses: unexpected detection of human herpesvirus 1 while searching for RNA pathogens. J Virol Methods 226:1–6

Quinones-Mateu ME, Avila S, Reyes-Teran G, Martinez MA (2014) Deep sequencing: becoming a critical tool in clinical virology. J Clin Virol 61(1):9–19

Rasool V, Rasool S, Mushtaq S (2014) Viral encephalitis and its management through advanced molecular diagnostic methods: a review. Clin Pediatr (Phila) 53 (2):118–120

Salter SJ, Cox MJ, Turek EM, Calus ST, Cookson WO, Moffatt MF, Turner P, Parkhill J, Loman NJ, Walker AW (2014) Reagent and laboratory contamination can critically impact sequence-based microbiome analyses. BMC Biol 12:87

Shen H, Rogelj S, Kieft TL (2006) Sensitive, real-time PCR detects low-levels of contamination by *Legionella pneumophila* in commercial reagents. Mol Cell Probes 20(3–4):147–153

Silva MT (2013) Viral encephalitis. Arq Neuropsiquiatr 71(9B):703–709

Solomon T, Hart IJ, Beeching NJ (2007) Viral encephalitis: a clinician's guide. Pract Neurol 7(5):288–305

Steiner I, Budka H, Chaudhuri A, Koskiniemi M, Sainio K, Salonen O, Kennedy PG (2005) Viral encephalitis: a review of diagnostic methods and guidelines for management. Eur J Neurol 12 (5):331–343

Strong MJ, Xu G, Morici L, Splinter Bon-Durant S, Baddoo M, Lin Z, Fewell C, Taylor CM, Flemington EK (2014) Microbial contamination in next generation sequencing: implications for sequence-based analysis of clinical samples. PLoS Pathog 10(11): e1004437

Tan le V, van Doorn HR, Nghia HD, Chau TT, le TP T, de Vries M, Canuti M, Deijs M, Jebbink MF, Baker S, Bryant JE, Tham NT NTBK, Boni MF, Loi TQ, Phuong le T, Verhoeven JT, Crusat M, Jeeninga RE, Schultsz C, Chau NV, Hien TT, van der Hoek L, Farrar J, de Jong MD (2013) Identification of a new cyclovirus in cerebrospinal fluid of patients with acute central nervous system infections. MBio 4(3):e00231–e00313

Towner JS, Sealy TK, Khristova ML, Albarino CG, Conlan S, Reeder SA, Quan PL, Lipkin WI, Downing R, Tappero JW, Okware S, Lutwama J, Bakamutumaho B, Kayiwa J, Comer JA, Rollin PE, Ksiazek TG, Nichol ST (2008) Newly discovered ebola virus associated with hemorrhagic fever outbreak in Uganda. PLoS Pathog 4(11):e1000212

Virgin HW, Todd JA (2011) Metagenomics and personalized medicine. Cell 147(1):44–56

Wilson MR, Naccache SN, Samayoa E, Biagtan M, Bashir H, Yu G, Salamat SM, Somasekar S, Federman S, Miller S, Sokolic R, Garabedian E, Candotti F, Buckley RH, Reed KD, Meyer TL, Seroogy CM, Galloway R, Henderson SL, Gern JE, DeRisi JL, Chiu CY (2014) Actionable diagnosis of neuroleptospirosis by next-generation sequencing. N Engl J Med 370(25):2408–2417

Yozwiak NL, Skewes-Cox P, Stenglein MD, Balmaseda A, Harris E, DeRisi JL (2012) Virus identification in unknown tropical febrile illness cases using deep sequencing. PLoS Negl Trop Dis 6(2): e1485

Advs Exp. Medicine, Biology - Neuroscience and Respiration (2017) 27: 63–71
DOI 10.1007/5584_2016_61

Published online: 29 July 2016

Mandibular Advancement Appliance for Obstructive Sleep Apnea Treatment

J. Kostrzewa-Janicka, P. Śliwiński, M. Wojda, D. Rolski, and E. Mierzwińska-Nastalska

Abstract

A combination of abnormal anatomy and physiology of the upper airway can produce its repetitive narrowing during sleep, resulting in obstructive sleep apnea (OSA). Treatment of sleep-breathing disorder ranges from lifestyle modifications, upper airway surgery, continuous positive airway pressure (CPAP) to the use of oral appliances. A proper treatment selection should be preceded by thorough clinical and instrumental examinations. The type and number of specific oral appliances are still growing. The mandibular advancement appliance (MAA) is the most common type of a dental device in use today. The device makes the mandible protrude forward, preventing or minimizing the upper airway collapse during sleep. A significant variability in the patients' response to treatment has been observed, which can be explained by the severity of sleep apnea at baseline and duration of treatment. In some trials, patients with mild OSA show a similar treatment effect after the use of CPAP or MAA. It is worthwhile to give it a try with an oral appliance of MAA type in snoring, mild-to-moderate sleep apnea, and in individuals who are intolerant to CPAP treatment.

Keywords

Airway collapse • Life style • Mandibular advancement device • Sleep apnea • Sleep disordered breathing • Treatment effectiveness • Upper airways

The original version of this chapter has been revised. An erratum to this chapter can be found at DOI 10.1007/978-3-319-44488-8_11

J. Kostrzewa-Janicka (✉), M. Wojda, D. Rolski, and E. Mierzwińska-Nastalska
Department of Prosthodontics, Medical University of Warsaw, Warsaw, Poland
e-mail: jolanta.kostrzewa-janicka@wum.edu.pl

P. Śliwiński
IV Fourth Clinic of Lung Diseases, Institute of Tuberculosis and Lung Diseases, Warsaw, Poland

1 Etiology and Epidemiology Obstructive Sleep Apnea (OSA)

OSA represents a group of sleep-breathing disorders and makes up 90 % of all forms of apnea syndromes (Hudgel 1992). It is characterized by the occurrence of upper airway obstruction during sleep at the pharyngeal level with intensified activity of respiratory muscles. Respiratory rhythm slows down during sleep and a decrease in muscular tension of soft palate muscles, along with the uvula, tongue, and the posterior pharyngeal wall, occurs. This results in retraction of the tongue and approaching of pharyngeal walls. In addition, gravitation in the supine position during sleep changes the position of the mandible, further decreasing in the pharyngeal space, especially in obese patients with accumulated adipose tissues in the submandibular region. The OSA-predisposing risk factors comprise obesity, mainly the deposition of adipose tissues in the neck, short and large neck, the structure of and anatomical changes in, the upper airways and craniofacial region (long soft palate, palatine adenoid tonsils, nasal polyps, nasal septum bending, long uvula, large tongue, elongated hard palate, maxillary protraction, body shortening and elongation of maxillary ramus, increment in mandibular angle, and excessive curvature of the thoracic spine), and the use of stimulants and medications reducing muscle tension (De Backer 2013; Sutherland et al. 2012; deBerry-Bobrowiecki et al. 1998; Lowe et al. 1986).

The adult incidence of OSA ranges from 3 to 7 %, depending on the gender and the subgroup characterized by different risk of this medical condition (Lurie 2011; Punjabi 2008). Longitudinal cohort studies show that a 10 % increase in body weight causes a 6-fold increase in the risk of moderate-to-severe OSA. A 5-year follow up study has shown that gender plays a significant role in the occurrence of OSA symptoms in middle-age patients (11.1 % in men and 5.9 % in women). Men with a 10 kg increase in body weight show a 5.2-fold risk of higher apnea/hypopnea index (AHI), over 15 events per hour, while in women this risk is a 2.5-fold higher.

2 OSA Diagnosis

According to the American Academy of Sleep Medicine Task Force Report (AASM 1999), OSA is recognized on the basis of the following specific diagnostic criteria: at least 5 events of obstruction of the upper airways per hour of sleep (AHI >5), excessive daytime sleepiness or at least two other symptoms, such as interrupted breathing or suffocation during sleep, recurrent arousal from sleep, sleep providing no relaxation, daily fatigue, and concentration difficulty. Three types of OSA have been identified: mild – characterized by the occurrence of fewer than 15 apnea/hypopnea events during sleep (AHI <15) and the patient's involuntary sleepiness during activities that require little attention; moderate – the number of apnea/hypopnea events ranges between 15 and 30 (AHI – 5–30) per hour and involuntary sleepiness occurs in activities that require some attention; and severe OSA – when over 30 apnea/hypopnea events per hour of sleep occur (AHI >30) and involuntary sleepiness during activities that require more active attention during daytime.

Polysomnography is the gold standard test for OSA diagnosis due to its ability to identify parameters described in the OSA definition. The excessive daytime sleepiness (EDS) is routinely measured by the Epworth Sleepiness Scale (ESS) (Johns 1991). If 10 OSA events occur per hour of sleep and persists each time longer than 10 s then OSA can be diagnosed.

3 OSA Treatment

As mentioned above, surgical and conservative techniques find their application in OSA treatment. Life style changes, body mass reduction, giving up stimulants and medication that has an effect on breathing (behavioral treatment) play a significant role in the OSA treatment (Campbell et al. 2015; Sharples et al. 2016; Carra et al. 2012; Hoffstein 2007; Padma et al. 2007). Surgical treatment is recommended only in cases of abnormal anatomy predisposing to the development of OSA and in hypertrophic changes. Of the surgical techniques applied, the following are worthwhile to emphasize: uvulopalatopharyngoplastics, i.e., techniques involving a partial resection of the soft palate, uvula, and palatal arches;

reduction of tongue volume and the correction of its position; forward movement of the maxilla, the mandible, and the hyoid bone using maxillary osteotomy of type Lefort I; and bilateral osteotomy of the mandible. Surgical removal of the accumulated adipose tissues in the region of chin and in other areas has also been applied (Padma et al. 2007).

Continuous positive air pressure (CPAP) is the most preferred conservative apnea treatment. It provides a steady stream of pressurized air to the airways using a soft plastic bulb connected with a mask covering a patient's nose or nose and mouth. Additional pressure ensures a mechanical stiffening of the upper airways protecting from the collapse of their walls.

4 Mandibular Advancement Appliance

Difficulties with adherence of patients to CPAP therapy and as a growing interest of physicians in the apnea phenomenon have encouraged further research to develop new therapeutic approaches, with the goal to expand the upper airway volume through pulling the lingual epiphysis away or mandibular advancement (Campbell et al. 2015; Hoffstein 2007; Schmidt-Nawara et al. 1995). A monoblock causing the mandible to protrude was first used by Pierre Robin in children with micrognation (Robin 1934). Two types of oral appliances to treat OSA are currently in use: tongue retaining device and mandibular advancement appliance (MAA) (Standards of Practice Committee of American Sleep Disorders Association 1995). The tongue retaining device, described by Cartwright (1985), protects against the tongue retraction. This action is possible due to hypotension retaining the tip of the tongue in the anterior part of the device. The device is constructed of a plastic bulb with a tongue-like shape. The tongue placed and sucked into the device is advanced forward so that the anterior-posterior size of the throat is expanded. In addition, tongue retaining device enhances the tension of the tongue and chin muscles. Nonetheless, it is used rather infrequently, due to certain discomfort it produces, although the device is very beneficial in edentulous patients.

The AASM has specified the indications for using dental appliances in patients with primary snoring without OSA; in mild OSA with reduced sleep apnea risk factors; and in moderate-to-severe OSA in patients who are intolerant of CPAP, who refused this kind of therapy, or in those not classified for surgery (Standards of Practice Committee of American Sleep Disorders Association 1995). Patients' eligibility for treatment with prosthetic devices should be preceded by a thorough polysomnographic, and intra- and extra-oral examinations, enabling the assessment of a possible application of MAA. The assessment takes into account such factors as the number and quality of preserved teeth and the condition of perodontium and temporomandibular joints. An attempt to treat OSA with the use of MAA should meet two major conditions, namely the presence of at least eight stable teeth in the maxilla and the mandible, and the possibility of establishing a constructive occlusion with the mandibular position in 50–75 % of maximum protrusion, leaving 3–5 mm of space between incisors allowing the patient to breathe freely through the mouth (Johal and Bottegal 2001). In edentulous patients, it is recommended to place the implants in the mandible for MAA fixing.

5 Methodology of Review

We set out to assess treatment effectiveness of MAA in patients with OSA as compared with placebo, conservative treatment, and CPAP. The assessment was based on the analysis of changes in AHI and ESS, and other commonly used indicators of the assessment of treatment effectiveness. The English-language literature, covering the period of 10 years, from 2004 to 2014, was retrieved in the Medline/PubMed database, using the query "obstructive sleep apnea/hypopnea" AND "mandibular advancement appliance" AND "continuous positive airway pressure". The adopted inclusion criteria consisted of the evidence of OSA in a polysomnographic examination and the randomized clinical trials comparing oral appliance with controls or other treatments in adults. Review papers, non-randomized trials, and articles without selected control groups and uniform diagnostic methods which could provide basis for diagnosing sleep apnea were excluded.

6 Results

We identified 59 clinical trials articles on the study subject published from the year 2000. Based on the inclusion and exclusion criteria for the studies assessing patients with OSA above outlined, 15 papers were selected, of which 10 compared therapeutic efficacy of MAA with CPAP (Tables 1 and 2) and

Table 1 Apnea-hypopnea index (AHI) after the use of different modes of treatment in obstructive sleep apnea (OSA) patients in randomized clinical trials

Study	Baseline	CPAP	MAA	Placebo
Aarab et al. (2011) (PSG evaluation)	Oral appliance: 21.4 ± 11.0 CPAP: 20.8 ± 9.9	Δ 20.2 ± 8.6$^{\wedge}$	Δ 15.0 ± 10.5	–
Barnes et al. (2004) (PSG evaluation)	21.3 ± 1.3	4.8 ± 0.5*†‡	14.0 ± 1.1*†	20.3 ± 1.1
Doff et al. (2013) (PSG evaluation)	Oral appliance: 39 ± 31 CPAP: 40 ± 28	0 (0–1)‡ (1-yr follow-up) 0 (0–1)$^{\#}$ (2-yr follow-up)	2 (0–5)^{ns} (1-yr follow-up) 2 (1–8)$^{\#}$ (2-yr follow-up)	–
Engelman et.al (2002) (home sleep study)	Oral appliance: 30 ± 21 CPAP: 32 ± 29	8 ± 6$^{\wedge}$	15 ± 16	–
Hoekema et al. (2008) (PSG evaluation)	Oral appliance: 39.4 ± 30.8 CPAP: 40.3 ± 27.6	7.8 ± 14.4$^{\beta}$	2.4 ± 4.2$^{\beta}$	–
Gagnadoux et al. (2009) (home sleep study)	34 ± 13	2 (1–8)*	6 (3–14)*	–
Lam et al. (2007) (PSG evaluation)	Oral appliance: 20.9 ± 1.7 CPAP: 23.8 ± 1.9	2.8 ± 1.1*	10.6 ± 1.7*	–
Phillips et al. (2013) (PSG evaluation)	25.6 ± 12.3	4.5 ± 6.6§	11.1 ± 12.1§	–
Randerath et al. (2002)	17.5 ± 7.7	3.5 ± 2.9§ (1 night) 3.2 ± 2.9$^{\S»}$ (6 weeks)	10.5 ± 7.5$^{\#}$ (1 night) 13.8 ± 11.1 (6 weeks)	–
Tan et al. (2002) (PSG evaluation)	22.2 ± 9.6	3.1 ± 2.8*	8.0 ± 10.9*	–

Data are means ± SD, values with additives in parenthesis are medians and interquartile ranges; Δ downward difference between baseline and treatment result, CPAP continuous positive airway pressure, MAA mandibular advancement appliance

$^{*}p < 0.001$ *vs.* baseline
$^{\dagger}p < 0.001$ *vs.* placebo
$^{\ddagger}p < 0.05$ *vs.* MAA
$^{\S}p < 0.01$ *vs.* baseline
$^{\#}p < 0.05$ *vs.* baseline
$^{»}p < 0.01$ *vs.* MAA
$^{\beta}p < 0.006$ *vs.* baseline
$^{\Delta}p < 0.001$ *vs.* MAA
^{ns}non-significant

Table 2 Epworth Sleepness Scale (ESS) or excessive daytime sleepiness (EDS) score after the use of different modes of treatment in obstructive sleep apnea (OSA) patients in randomized clinical trials

Study	Baseline	CPAP	MAA	Placebo
Aarab et al. (2011) EDS	Oral appliance:12.0 ± 5.7 CPAP: 11.0 ± 4.4	Δ 5.2 ± 4.6^	Δ 4.7 ± 4.5	–
Barnes et al. (2004) ESS	10.7 ± 0.4	9.2 ± 0.4*†	9.2 ± 0.4*†	10.2 ± 0.4
Doff et al. (2013) ESS	Oral appliance: 13 ± 6 CPAP: 14 ± 6	5 (2–9)ns (1-yr follow-up) 5 (1–8)# (2-yr follow-up)	5 (3–8)ns (1-yr follow-up) 4 (1–8)# (2-yr follow-up)	–
Engelman et.al (2002) ESS	14 ± 4	8 ± 5^	12 ± 5	–
Hoekema et al. (2008) (PSG evaluation)	Oral appliance: 12.5 ± 5.6 CPAP: 14.2 ± 5.6	6.9 ± 5.5ns	5.9 ± 4.8ns	–
Gagnadoux et al. (2009) ESS	10.6 ± 4.5	8.2 ± 3.9*	7.7 ± 4.0*	–
Lam et al. (2007) ESS	Oral appliance: 20.0 ± 1.7 CPAP: 23.8 ± 1.9	2.8 ± 11.0*	10.6 ± 1.7*	–
Phillips et al. (2013) ESS	9.1 ± 4.2	7.5 ± 0.4§	7.2 ± 0.4§	–
Tan et al. (2002) ESS	13.4 ± 4.6	8.1 ± 4.1*	9.0 ± 5.1*	–

Data are means ± SD, values with additives in parenthesis are medians and interquartile ranges; ΔEDS downward difference between baseline and treatment result, CPAP continuous positive airway pressure, MAA mandibular advancement appliance

*p < 0.001*vs.* baseline

†p < 0.001 *vs.* placebo

§p < 0.01 *vs.* baseline

#p < 0.05 *vs.* baseline

^p < 0.001 *vs.* MAA

nsnon-significant

5 compared therapeutic efficacy of different types of MAA with controls or placebo (Quinnell et al. 2014; Petri et al. 2008; Blanco et al. 2005; Gotsopoulos et al. 2002; Mehta et al. 2001).

The papers that contain a description of baseline characteristics indicated the beginning of treatment in moderate and severe cases of OSA. The AHI value after application of both CPAP and MAA decreases to the level of mild OSA (Doff et al. 2013; Aarab et al. 2011; Gagnadoux et al.. 2009; Hoekema et al. 2008; Lam et al. 2007; Engelman et al. 2002; Tan et al. 2002). However, AHI value after CPAP application is significantly lower than that after MAA application (Phillips et al. 2013; Barnes et al. 2004; Randerath et al. 2002). In the study of Aarab et al. (2011), the difference in AHI index between baseline and CPAP treatment is down to more than 20 while that between baseline and MAA treatment is 15 episodes per hour of sleep. Likewise, excessive daytime sleepiness appreciably decreased after both CPAP and MAA, with no significant difference between the two therapeutic modes.

All clinical trials perused in the present review indicate an improvement after treatment with both CPAP and MAA. However, if the treatment efficacy goal is set the AHI ≥ 5 apneic episodes per hour of sleep, then MAA usually fails to reach that cut-off level and thus should not rather be recommended as the first line treatment for OSA patients (Engelman et al. 2002).

On the other side, studies assessing the sustenance of improvement indicate a significant decrease in AHI compared with baseline over a two-year treatment time, using both methods (Doff et al. 2013). These results encourage to consider the use of oral appliance as a viable alternative to CPAP therapy in patients with mild-to-moderate OSA. However, CPAP remains undubiously the treatment of first choice.

Following the treatment with both CPAP and MAA, arterial blood oxygen saturation significantly increases; the increase is greater after CPAP, but MAA also significantly increases arterial blood oxygen saturation compared with placebo (Doff et al. 2013; Barnes et al 2004). Tan et al. (2002) have reported decreases in the mean arterial blood desaturation from the basal OSA level of 7.1 ± 2.7 % to 3.3 ± 1.6 % after CPAP and 4.8 ± 2.7 % after MAA; the former decrease being significant while the latter being insignificant.

The subjective evaluation of CPAP and MAA is distinctly different from the objective index-based evaluation. The ESS score decreases significantly in relation to baseline, with no significant differences between the CPAP and MAA effects (Table 2). The study of Mehta et al. (2001) performed in a group of 24 OSA patients has shown a good tolerance of MAA in 90 % of patients and the AHI decreased below 5 apneic episodes per hour in 37.5 % of patients. When the cut-off level for therapeutic efficacy of AHI is taken as fewer than 10 apneic episodes per hour, then the percentage of patients with adequate therapeutic response increased to 54 % compared with untreated OSA control subjects. Almost 70 % of patients reported a reduced snoring and in 91 % of patients the sleep quality improved. Similar findings, indicating a significant subjective improvement after MAA application, compared with controls using inactive oral appliances, have been presented by Gotsopoulos et al. (2002), even though objective improvements were lacking. Therapeutic efficacy of MAA is distinctly dependent on the degree of mandibular protrusion achieved with the appliance (Petri et al. 2008; Hans et al. 1997; Blanco et al. 2005).

Randomized studies aimed at the assessment of efficacy of different standard splints (self-molded, semi-bespoke, and fully-bespoke) for OSA treatment have also been performed and the findings compared with the observations in untreated patients (Quinnell et al. 2014). All these appliances decrease the AHI value by 26 % compared with no treatment (95 % CI; 11–38 %; $p = 0.001$). Further, the splints considerably improve the patients' self-assessment. Non-adjustable MAA achieve clinically important improvements in mild-to-moderate OSA and are cost-effective.

7 Synopsis

It is well known that the use of CPAP is the most effective protection from obstructive apneic episodes at sleep. Regular CPAP application causes a regression of apneic events, normalizes circadian disturbance of hormonal secretion, decreases blood pressure, and improves sleep quality. This method can be applied in edentulous patients. A complete nasal patency is the only requirement for the method. CPAP is a relatively safe procedure. However, its long-term application carries risk of certain complications; notably of nasal injuries such as necrosis, irritation and swelling of mucous membrane, or nasal septum bending. In 40 % of patients, upper respiratory complaints also occur such as rhinitis, sneezing, and mucosal drying. The accumulation of stomach gases, caused by swallowed air, is frequently reported. It also happens that the method itself proves ineffective, which leads to the development of atelectasis (Sharples et al. 2016; Phillips et al. 2013; Fergusson et al. 1996; Sallivan et al. 1981).

Bearing in mind the considerations above outlined, numerous research centers carry out reliable randomized studies, which create the basis for assessing how far other techniques could go to ensure safe sleep and optimal upper airway patency for adequate supply of oxygen.

Mandibular advancement appliances are designed to maintain the mandible in the protrusive position, to dislocate forward the tongue *via* the genioglossus muscles and to change the position of hyoid bone, thus expanding the upper airways (Franson et al. 2002). It has been found that the application of MAA only to stimulate the genioglossus muscles without mandibular advancement has no influence on the number of obstructive respiratory events during sleep (Metha et al. 2001). Upon establishing an optimal mandibular position with the appliance to reach the therapeutic goal, a control polysomnographic examination should be performed to assess the device efficacy, since in some patients MAA may inadvertently increase the number of apneic events (Ferguson et al. 1997). Nevertheless, numerous findings confirm the overall effectiveness of mandibular advancement devices for OSA treatment. The results of magnetic resonance imaging and endoscopic examinations provide evidence that MAA application expands the volume of upper airways (Gao et al. 1999).

Hoffstein (2007) has divided the adverse effects of mandibular advancement appliances into three groups: mild-transient, moderate-to-severe, and continuous. Excess salivation, dry mouth, allergic reactions to the applied materials, and pain in temporomandibular joints are the adverse effects observed most frequently. The available literature reports that cephalometric findings and model analyses reveal changes in the overjet and overbite, maloclusion, changes in the upper incisor angle to cranial base (1/NS) and the angle between Sella-Nasion-Supramentale point after MAA application (Fritsch et al. 2001). In a study carried out to assess the effects of the Herbst mandibular advancement appliances employed for two years, Battagel and Kotecha (2005) have observed changes in the position of incisors, insignificantly diminished overjet, and the overbite correlation with the splint vertical dimension. These changes were related neither to the degree of mandibular advancement nor the period of device use. Fransson et al. (2003) have reported similar findings. Investigating the effect of MAA on the stomatognathic system, Rinqvist et al. (2003) have used splints that do not cover the anterior region of the dental arch, with the mandibular advancement not exceeding 50 %. The authors did not notice dental abnormalities in the form of changed overjet, overbite, or inclination angle of upper and lower incisors. Marklund et al. (2004) have rarely observed adverse effects with the use of devices produced of soft material. On the other hand, Bondemark and Lindman (2000) have found that devices made of hard material better prevent upper airway occlusion, giving a full support to dental arches. Martinez-Gomis et al. (2010) have shown that the majority of dental changes occur during the first two years of using mandibular advancement appliances.

8 Conclusions

The available research demonstrates that treatment of obstructive sleep apnea with mandibular advancement appliances is well grounded. The use of such appliances is an effective therapeutic alternative in carefully selected clinical forms of OSA and may give an edge over the standard CPAP treatment. Mandibular advancement appliances seem a particularly attractive therapeutic alternative in mild-to-moderate OSA and in CPAP intolerant patients. These appliances also have a distinct reducing effect on the overwhelming feeling of daily sleepiness. It should be borne in mind that OSA is a disease of many faces, due to, among others, different severity of disease at onset, and differences in morphology and duration of treatment. The response to treatment may thus be highly and unpredictably variable.

Conflicts of Interest The authors declare no conflicts of interest in relation to this article.

References

Aarab G, Lobbezzo F, Hamburger HL, Naeije M (2011) Oral appliance therapy versus nasal continuous positive airway pressure in obstructive sleep apnea; a randomized, placebo-controlled trial. Respiration 81:411–419

AASM (1999) American Academy of Sleep Medicine Task Force Report. Sleep-related breathing disorders in adults: recommendations for syndrome definition and measurement techniques in clinical research. Sleep 22(5):667–689

Fransson AM, Tegelberg A, Leissner L, Wenneberg B, Isacsson G (2003) Effects of a mandibular protruding device on the sleep of patients with obstructive sleep apnea and snoring problems: a 2-year follow-up. Sleep Breath 7(3):131–141

Barnes MR, McEvoy D, Banks S, Tarquinio N, Murray CG, Vowles N, Pierce RJ (2004) Efficacy of positive airway pressure and oral appliance in mild to moderate obstructive sleep apnea. Am J Respir Crit Care Med 170:656–664

Battagel JM, Kotecha B (2005) Dental side effects of mandibular advancement splint wear in patients who snore. Clin Otolaryngol 30:146–156

Blanco J, Zamarron C, Abeleira Pazos MT, Lamela C, Suarez Quintanilla D (2005) Prospective evaluation of an oral appliance in the treatment of obstructive sleep apnea syndrome. Sleep Breath 9:20–25

Bondemark L, Lindman R (2000) Craniomandibular status and function in patients with habitual snoring and obstructive sleep apnea after nocturnal treatment with a mandibular advancement splint: a 2-year follow-up. Eur J Orthod 22(1):53–60

Campbell T, Pengo MF, Steier J (2015) Patients preference of established and emerging treatment options for obstructive sleep apnoea. J Thorac Dis 7(5):938–942

Carra MC, Bruni O, Huynh N (2012) Topical review: sleep bruxism, headaches, and sleep-disordered breathing in children and adolescents. J Orofac Pain 26:267–276

Cartwright RD (1985) Predicting response to the tongue retaining device for sleep apnea syndrome. Arch Otolaryngol 111(6):385–388

De Backer W (2013) Obstructive sleep apnea/hypopnea syndrome. Panminerva Med 55(2):191–195

deBerry-Bobrowiecki B, Kukwa A, Blanks RH (1998) Cephalometric analysis for diagnosis and treatment of obstructive sleep apnea. Laryngoscope 98(2):226–234

Doff MH, Hoekema A, Wijkstra PJ, van der Hoeven JH, Huddleston Slater JJR, de Bont LGM, Stegenga B (2013) Oral appliance versus continuous positive airway pressure in obstructive sleep apnea syndrome: a 2-year follow-up. Sleep 36(9):1289–1296

Engelman HM, McDonald JP, Graham D, Lello GE, Kingshott RN, Coleman EL, Mackay TW, Douglas NJ (2002) Randomized crossover trial of two treatments for sleep apnea/hypopnea syndrome: continuous positive airway pressure and mandibular repositioning splint. Am J Respir Crit Care Med 166:855–859

Ferguson KA, Ono T, Lowe AA, Al- Majed S, Love LL, Fleetham JA (1997) A short-term controlled trial of an adjustable oral appliance for the treatment of mild to moderate obstructive sleep apnea. Thorax 52:362–368

Fergusson KA, Ono T, Lowe AA, Keenan SP, Fleetham JA (1996) A randomized crossover study of an oral appliance versus nasal continuous positive airway pressure in the treatment of mild-moderate sleep apnea. Chest 109:1269–1275

Franson AMC, Tegelberg Å, Svenson BAH, Lennartsson B, Isacsson G (2002) Influence of mandibular protruding device on airway passages and dentofacial characteristcs in obstructive sleep apnea and snoring. Am J Orthod Dentofacial Orthop 122(4):371–379

Fritsch KM, Iseli A, Russi EW, Bloch KE (2001) Side effects of mandibular advancement devices for sleep apnea treatment. Am J Respir Crit Care Med 164 (5):813–818

Gagnadoux F, Fleury B, Vielle B, Petelle B, Meslier N, N'Guyen XL, Trzepizur W, Racineux JL (2009) Titrated mandibular advancement versus positive airway pressure for sleep apnoea. Eur Respir J 34:914–920

Gao XM, Zeng XL, Fu MK, Huang XZ (1999) Magnetic resonance imaging of the upper airway in obstructive sleep apnea before and after oral appliance therapy. Chin J Dent Res 2(2):27–35

Gotsopoulos H, Chen C, Quian J, Cistulli PA (2002) Oral appliance therapy improves symptoms in obstructive sleep apnea. Am J Respir Crit Care Med 166:743–748

Hans MG, Nelson S, Luks VG, Lorkovich P, Baek SJ (1997) Comparison of two dental devices for treatment of obstructive sleep apnea syndrome (OSAS). Am J Orthod Dentofacial Orthop 111:562–570

Hoekema A, Stegenda B, Wijkstra PJ, van der Hoeven JH, de Bont LG (2008) Obstructive sleep apnea therapy. J Dent Res 87(9):882–887

Hoffstein V (2007) Review of oral appliance for treatment of sleep-disordered breathing. Sleep Breath 11:1–22

Hudgel DW (1992) Mechanism of obstructive sleep apnea. Chest 101(2):541–549

Johal A, Bottegal JM (2001) Current principles in the management of obstructive sleep apnea with mandibular advancement appliances. Br Dent J 190(100):532–536

Johns MW (1991) A new method for measuring daytime sleepiness: the Epworth sleepiness scale. Sleep 31:1551–1558

Lam B, Sam K, Mok WY, Cheung MT, Fong DY, Lam JC, Lam DC, Yam LY, Ip MS (2007) Randomised study of three non-surgical treatments in mild to moderate obstructive sleep apnoea. Thorax 62:354–359

Lowe AA, Santamaria JD, Fleetham JA, Price C (1986) Facial morphology and obstructive sleep apnea. Am J Orthod Dentofacial Orthop 90(1):484–491

Lurie A (2011) Obstructive sleep apnea in adults: epidemiology, clinical presentation, and treatment options. Adv Cardiol 46:1–42

Marklund M, Stenlund H, Franklin KA (2004) Mandibular advancement devices in 630 men and women with obstructive sleep apnea and snoring: tolerability and predictors of treatment success. Chest 125(4):1270–1272

Martinez-Gomis J, Willaert E, Nogues L, Pascual M, Somoza M, Monasterio C (2010) Five years of sleep apnea treatment with a mandibular advancement device. Side effects and technical complications. Angle Orthod 80:30–36

Mehta A, Qian J, Petocz P, Darendeliler MA, Cistulli PA (2001) A randomized, controlled study of a mandibular advancement splint for obstructive sleep apnea. Am J Respir Crit Care Med 163:1457–1461

Padma A, Ramakrishnan N, Narayanan V (2007) Management of obstructive sleep apnea: a dental perspective. Indian J Dent Res 18(4):201–209

Petri N, Svanholt P, Solow B, Wildschiodtz G, Winkel P (2008) Mandibular advancement appliance for obstructive sleep apnoea: results of a randomised placebo controlled trial using parallel group design. J Sleep Res 17:221–229

Phillips CL, Grunstein RR, Darendeliler MA, Mihailidou AS, Srinivasan VK, Yee BJ, Marks GB, Cistulli PA (2013) Health outcomes of continuous positive airway pressure versus oral appliance treatment for obstructive sleep apnea: a randomized controlled trial. Am J Respir Crit Care Med 187(8):879–887

Punjabi NN (2008) The epidemiology of adult obstructive sleep apnea. Proc Am Thorac Soc 5(2):136–143

Quinnell TG, Bennett M, Jordan J, Clutterbuck-James AL, Davies MG, Smith IE, Oscroft N, Pittman MA, Cameron M, Chadwick R, Morrell MJ, Glover MJ, Fox-Rushby JA, Sharples LD (2014) A crossover randomised controlled trial of oral mandibular advancement devices for obstructive sleep apnoea-hypopnoea (TOMADO). Thorax 69:938–945

Randerath WJ, Heise M, Hinz R, Ruehle KH (2002) An individually adjustable oral appliance vs continuous positive airway pressure in mild –to-moderate obstructive sleep apnea syndrome. Chest 122:569–575

Rinqvist M, Walker-Engstrom M, Tegelberg A, Rinqvist I (2003) Dental and skeletal changes after 4 years of obstructive sleep apnea treatment with a mandibular advancement device: a prospective, randomized study. Am J Orthod Dentofacial Orthop 124:53–60

Robin P (1934) Glossoptosis due to artresia and hypotrophy of the mandible. JAMA Pediatr 48 (3):541–547

Sallivan CE, Issa FG, Berghton-Jones M, Eves L (1981) Reversal of obstructive sleep apnea by continuous positive airway pressure applied through the noses. Lancet 1(8225):862–865

Schmidt-Nawara W, Lowe A, Wiegand L, Cartwright R, Perez-Gguerra F, Menn S (1995) Oral appliances for the treatment of snoring and obstructive sleep apnea: a review. Sleep 18(6):501–510

Sharples LD, Clutterbuck-James AL, Glover MJ, Bennett MS, Chadwick R, Pittman MA, Quinnell TG (2016) Meta-analysis of randomised controlled trials of oral mandibular advancement devices and continuous positive airway pressure for obstructive sleep apnoea-hypopnoea. Sleep Med Rev 27:108–124

Standards of Practice Committee of American Sleep Disorders Association (1995) Practice parameters for the treatment of snoring and obstructive sleep apnea with oral appliance. Sleep 18:511–513

Sutherland K, Lee RW, Cistulli PA (2012) Obesity and craniofacial structure as risk factors for obstructive sleep apnea: impact of ethnicity. Respirology 17(2):213–222

Tan YK, L'Estrange PR, Luo YM, Smith C, Grant HR, Simonds AK, Spiro SG, Battagel JM (2002) Mandibular advancement splints and continuous positive airway pressure in patients with obstructive sleep apnoea: a randomized cross-over trial. Eur J Orthod 24:239–249

Advs Exp. Medicine, Biology - Neuroscience and Respiration (2017) 27: 73–80
DOI 10.1007/5584_2016_43

Published online: 13 July 2016

Acute Response to Cigarette Smoking Assessed in Exhaled Breath Condensate in Patients with Chronic Obstructive Pulmonary Disease and Healthy Smokers

M. Maskey-Warzęchowska, P. Nejman-Gryz, K. Osinka, P. Lis, K. Malesa, K. Górska, and R. Krenke

Abstract

The effect of acute exposure to cigarette smoke (CS) on the respiratory system has been less extensively studied than the long term effects of smoking. The aim of the present study was to evaluate the acute response to CS in smokers suffering from chronic obstructive pulmonary disease (COPD) and in healthy smokers. Nineteen stable COPD patients and 19 young healthy smokers were enrolled. Tumor necrosis factor alpha (TNF-α), IL-1β, and malondialdehyde (MDA) were measured in exhaled breath condensate (EBC) before and 60 min after smoking a cigarette. When pre- and post-CS levels of the evaluated biomarkers were compared, no differences were found in either group. However, the post-CS MDA was significantly greater in healthy smokers than that in COPD patients; 20.41 vs. 16.81 nmol/L, $p = 0.01$, respectively. Post-CS TNF-α correlated inversely with FEV_1/FVC in healthy smokers. We conclude that CS does not acutely increase the EBC concentration of the inflammatory markers either in COPD patients or healthy smokers. The short term CS-induced oxidative stress is higher in young smokers than in COPD patients, which what may indicate a higher susceptibility to CS content of the former.

Keywords

Airway obstruction • Inflammation • Malondialdehyde • Smoking • Tumor necrosis factor alpha

The original version of this chapter has been revised. An erratum to this chapter can be found at DOI 10.1007/978-3-319-44488-8_11

M. Maskey-Warzęchowska, P. Nejman-Gryz, K. Górska (✉), and R. Krenke
Department of Internal Medicine, Pneumology and Allergology, Medical University of Warsaw, 1A Banacha, 02-097 Warsaw, Poland
e-mail: drkpgorska@gmail.com

K. Osinka, P. Lis, and K. Malesa
Student Scientific Association 'Alveolus', Department of Internal Medicine, Pneumology and Allergology, Medical University of Warsaw, Warsaw, Poland

1 Introduction

Cigarette smoking is responsible for 6 million deaths each year (WHO 2016). The 2012 data indicate that 21 % of the world population aged 15 and more smoked. It is estimated that approximately 32.4 % of men and 23.7 % of women were smokers in Poland in 2015 (WHO 2016). The impact of smoking on general health is well documented. Despite numerous anti-tobacco campaigns, prevalence of smoking-related diseases is alarming. Chronic obstructive pulmonary disease (COPD), for which smoking is the major risk factor, was the world's sixth leading cause of death in 1990 and is presumed to rank fourth by 2030 (GOLD 2015). Smoking is also the main risk factor for lung cancer, a leading cancer worldwide in both genders (WCRF 2016).

COPD and lung cancer usually develop after many years of exposure to tobacco, therefore the vast majority of studies focus on the long-term effect of smoking. Exposure to cigarette smoke (CS) increases the production of a variety of compounds which trigger a complex inflammatory response leading to structural changes in the respiratory tract. The proinflammatory compounds include tumor necrosis factor α (TNF-α), interleukin 1β (IL-1β), granulocyte-macrophage colony stimulating factor (GM-CSF), interleukin 8 (IL-8), transforming growth factor β (TGF-β), metalloproteinases, cathepsins, neutrophil elastase, and reactive oxygen species (Rovina et al. 2013).

Research on the acute effect of cigarette smoke is less extensive. Smoking is followed by instant changes the physico-chemical properties of exhaled breath condensate causing an increase in its electric conductivity (Koczulla et al. 2010), changes in pH, in markers of oxidative stress, and in the cytokine and chemokine profiles (Konstantinidi et al. 2015). It has been documented that acute exposure to CS stimulates the release of IL-1β and TNF-α by peripheral blood mononuclear cells (Ryder et al. 2002), increases the plasma level of malondialdehyde (MDA), a marker of oxidative stress, and reduces the plasma antioxidant potential (Durak et al. 2000). There is evidence that CS elicits the production of IL-1β by peripheral blood mononuclear cells in *in vitro* (Ryder et al. 2002). CS also causes short-term alterations in the cellular composition of induced sputum in healthy intermittent smokers (van der Vaart et al. 2005). Studies involving bronchoalveolar lavage fluid have shown that CS stimulates influx of dendritic cells, mononuclear cells, and neutrophils to the airways (van der Toorn et al. 2013; Lommatzsch et al. 2010), and an increase of oxidative stress (van der Vaart et al. 2004). The evaluation of the inflammatory response to acute CS exposure has been performed at various time intervals, ranging from 5 to 15 min (Koczulla et al. 2010; Ryder et al. 2002) to 8 days (van der Vaart et al. 2005) after smoking. Some studies consist of serial measurements (Koczulla et al. 2010; van der Vaart et al. 2005).

Given different biologic samples tested and different time points at which the effect of short term smoking on the airways has been investigated, we undertook this study to evaluate the level of TNF-α, IL-1β, and MDA in exhaled breath condensate (EBC), as measures of inflammation and oxidative stress burden, in response to acute CS exposure, i.e., 60 min after smoking a single cigarette, in smokers suffering from COPD and in healthy smokers.

2 Methods

2.1 Study Design

The study protocol was approved by the Bioethics Committee of the Medical University of Warsaw in Poland (permit #KB/106/2013). All the study participants signed informed consent. The study involved two groups: smoking patients with stable COPD and young healthy asymptomatic smokers. In each group, EBC was collected twice. The first sample was obtained in the morning after at least 6 h of refraining from smoking during nighttime. The second sample was collected 60 min after smoking a single cigarette of a commercial brand with a tar content of 6 mg and nicotine content of 0.5 mg. Both samples were analyzed for IL-1β, TNF-α, and MDA.

2.2 Study Participants

Nineteen consecutive active smokers with stable COPD who agreed to participate in the study during a routine follow-up visit in the outpatient clinic of the Department of Internal Medicine, Pneumology and Allergology of Medical University of Warsaw in Poland were enrolled into the study. The inclusion criteria were as follows: 1/diagnosis of COPD in accordance with the recommendations of the Global Initiative for Obstructive Lung Disease (GOLD 2015); 2/post-bronchodilator forced expiratory volume in one second (FEV_1) $\geq 50\ \%$ of predicted value; 3/current smoking; 4/lack of exacerbation or respiratory infection within 4 weeks before inclusion; and 5/no systemic or inhaled steroid use 6 weeks prior to study onset.

The second investigated group consisted of 19 asymptomatic young (<25 years of age) current smokers with a negative history of respiratory diseases and normal spirometry. Subjects who had a respiratory infection within 4 weeks before the study onset were excluded.

In all patients lung function was assessed by spirometry (Lungtest 1000, MES, Cracow, Poland), which was performed in accordance with the recommendations of the European Respiratory Society and the American Thoracic Society (Miller et al. 2005; Pellegrino et al. 2005) after refraining from smoking for at least 6 hours before the first EBC collection. In COPD patients, spirometry was performed after administration of their routine inhaled medication.

2.3 Exhaled Breath Condensate (EBC) Collection and the Evaluation of Biomarkers

EBC was collected and processed according to the ATS/ERS recommendations (Horvath et al. 2005) with the use of the TURBO-DECCS 09 system (Medivac, Parma, Italy) during tidal breathing for 20 min and −5 °C condensation temperature. The obtained samples were portioned in 250 μL aliquots and immediately stored at −70 °C for subsequent analysis.

The concentration of MDA, IL-1β, and TNF-α in EBC was measured by ELISA method. The following commercial ELISA kits were used: CLIA Kit for general malondialdehyde (EIAab; Wuhan, China, Catalog No EO597Ge), Human IL-1 beta ELISA kit (RayBiotech; Norcross, GA, Catalog No: ELH-IL1b), QuantikineHS ELISA Human TNF-alpha immunoassay (R&D; Minneapolis, MN, Catalog NoHSTA00D). The sensitivity of the applied kits was 0.15 nmol/mL for MDA, 0.3 pg/mL for IL-1β, and 0.191 pg/mL for TNF-α, respectively. EBC samples for IL-1β and TNF-α measurements were undiluted, while MDA levels were evaluated in twice diluted condensate samples. None of the samples were repeatedly thawed/frozen for analysis. To achieve maximal reproducibility and reliability of our measurements the samples were not lyophilized.

2.4 Statistical Analysis

Data are presented as median and interquartile range (IQR). The differences between continuous variables in unrelated groups were tested with the use the non-parametric Mann–Whitney U test. Spearman's rank correlation coefficient was applied to test potential correlations between different continuous variables. The differences between the related samples (repeated measurements) were tested with non-parametric

Wilcoxon signed-rank test. Statistical significance was accepted at a p-value less than 0.05. Analyses were performed using Statistica 10.0 (StatSoft Inc., Tulsa, OK) software package.

3 Results

3.1 Patient Characteristics

Basic characteristics of the study groups are listed in Table 1.

3.2 Baseline Level of TNF-α, IL-1β, and MDA in Exhaled Breath Condensate in COPD patients and Healthy Smokers

There were no differences between baseline EBC concentration of the evaluated biomarkers in COPD patients and healthy smokers (Fig. 1).

3.3 Effects of Smoking on Biomarkers in Exhaled Breath Condensate in COPD Patients and Healthy Smokers

There were no differences in the concentration of all three markers between their respective values at baseline and 60 min after smoking a single cigarette (Table 2).

The level of TNF-α after smoking was significantly lower in healthy smokers than that in patients with COPD (0.27 pg/mL vs. 0.46 pg/mL, $p = 0.035$). On the contrary, concentration of MDA after smoking was higher in healthy smokers compared with COPD patients (20.41 nmol/mL vs. 16.81 nmol/mL, $p = 0.01$) (Fig. 2).

3.4 Correlations Between EBC Biomarkers and Clinical Data

There was a negative correlation between baseline TNF-α concentration in EBC and duration of cigarette smoking in healthy smokers ($r = -0.56$, $p = 0.01$). Such a relationship was not confirmed in COPD patients. The level of TNF-α after smoking inversely correlated with FEV_1/FVC in the control group ($r = -0.59$, $p = 0.008$), but not in the group of COPD patients. In the COPD group, we found a relationship between baseline MDA concentration in EBC and FEV_1 and FVC (% of predicted value): $r = 0.52$, $p = 0.02$ and $r = 0.52$, $p = 0.02$, respectively. No other correlations between the EBC biomarkers and the investigated variables were demonstrated.

Table 1 Basic demographic and clinical data of the patients with chronic obstructive pulmonary disease (COPD) and healthy cigarette smokers participating in the study

	COPD patients n = 19	Healthy smokers n = 19	p
Gender (F/M)	8/11	8/11	ns
Age (year)	65 (61–69)	23 (21–24)	<0.0001
BMI (kg/m^2)	23.5 (21.2–27.8)	23.5 (21.1–26.3)	ns
Number of pack-years	33 (24–44)	2.1 (0.7–4)	<0.0001
Duration of smoking (year)	42 (35–47)	6 (5–7)	<0.0001
FEV_1 (L)	1.71 (1.29–2.41)	4.06 (3.4–4.6)	<0.0001
FEV_1 (% predicted)	61 (50–86)	101 (96–113)	<0.0001
FVC (L)	3.43 (2.32–3.99)	4.95 (3.90–5.80)	<0.0001
FVC (% predicted)	87 (81–109)	100 (93–108)	ns
FEV_1/FVC (%)	58 (47.5–60)	84 (81–90.5)	<0.0001

Data are medians and interquartile ranges (IQR). *BMI* body mass index, *FEV_1* forced expiratory volume in one second, *FVC* forced vital capacity, *ns* not significant

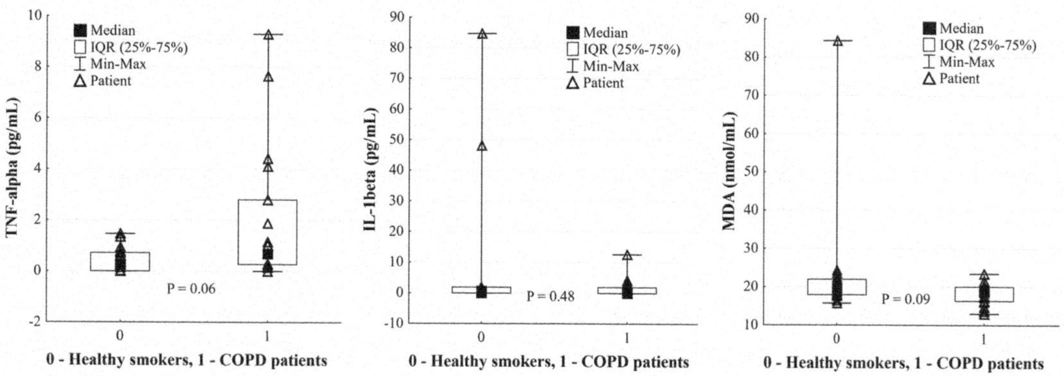

Fig. 1 Tumor necrosis factor alpha (TNF-α), interleukin 1beta (IL-1β), and malondialdehyde (MDA) in exhaled breath condensate (EBC) in healthy smokers and COPD patients at baseline (before smoking)

Table 2 IL-β, TNF-α, and MDA in exhaled breath condensate before and 60 min after smoking a single cigarette: comparison of COPD patients and healthy asymptomatic smokers

	COPD patients			Healthy smokers		
Biomarker	Baseline	60 min after smoking	p	Baseline	60 min after smoking	p
IL-1β (pg/mL)	0.37	0.74	ns	0.74	0.37	ns
	(0.0–1.85)	(0.00–2.59)		(0.00–1.85)	(0.00–3.33)	
TNF-α (pg/mL)	0.68	0.46	ns	0.27	0.27	ns
	(0.27–2.8)	(0.27–1.03)		(0.00–0.71)	(0.00–0.68)	
MDA (nmol/mL)	18.95	16.81	ns	19.74	20.41	ns
	(16.3–20.02)	(13.56–19.85)		(17.93–21.98)	(18.27–21.87)	

Data are medians and interquartile ranges (IQR). *IL*-1β interleukin 1β, *TNF*-α tumor necrosis factor α, *MDA* malondialdehyde, *ns* not significant

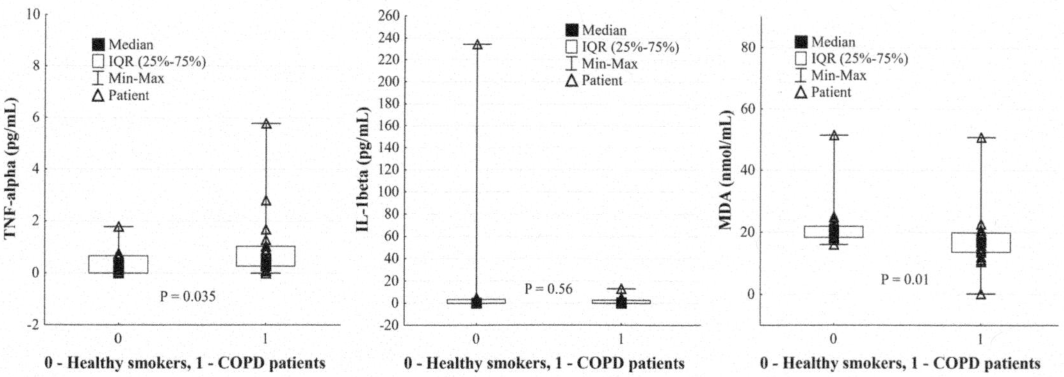

Fig. 2 Tumor necrosis factor alpha (TNF-α), interleukin 1beta (IL-1β), and malondialdehyde (MDA) in exhaled breath condensate (EBC) in healthy smokers and COPD patients 60 min after smoking a single cigarette

4 Discussion

In this study we assessed the impact of acute cigarette smoking on inflammatory cytokines and on the level of oxidative stress in EBC in smoking COPD patients and healthy asymptomatic smokers. We found that the level of TNF-α, IL-β, and MDA did not differ between both groups at baseline and did not change significantly 60 min after smoking a single cigarette.

There was a negative correlation between TNF-α concentration after smoking and FEV_1/FVC in healthy smokers, which may indicate that EBC TNF-α may be useful in identifying smoking subjects at risk for the development of COPD. Thus, we suppose that the result of our study may have practical implications in helping reduce the burden of COPD.

Acute pulmonary response to CS has been evaluated in a number of clinical studies. The protocols applied in earlier studies generally differ in three aspects: the magnitude of CS exposure (from a single cigarette to four cigarettes at a time), the time points at which the assessment was performed, and the biologic material used for investigation. CS-induced reactions have been assessed in isolated white blood cell populations (Hoonhorst et al. 2014; van der Vaart et al. 2004; Ryder et al. 2002), bronchoalveolar lavage (van der Toorn et al. 2013; Lommatzsch et al. 2010), bronchial biopsies (Hoonhorst et al. 2014), and induced sputum (Comandini et al. 2009). Over the past decade, there has been an increasing interest in exhaled breath condensate. Being non-invasive and relatively easily accessible, EBC is a particularly attractive material for the investigation of acute response to CS (Kostikas et al. 2013; Koczulla et al. 2010; Papaioannou et al. 2010; Garey et al. 2004). However, the use of EBC as a research tool has certain drawbacks. The main EBC component is water vapor, and therefore most biomarkers are usually detected at very low concentrations, at times very close to the their detection limit (Konstantinidi et al. 2015; Kotz et al. 2007; Horvath et al. 2005). Lyophilization may be applied to concentrate the sample, but little is known about the influence of this procedure on the composition of EBC and the potential impact on individual EBC biomarkers. In the present study EBC was not lyophilized. That may have contributed to the low levels of biomarkers investigated and the lack of differences between them before and after smoking. Corradi et al. (2003) have demonstrated that although a greater level of MDA in EBC discriminates smoking COPD patients and healthy smokers from non-smokers, MDA is not different between COPD and healthy smokers. This is in line with the present findings, which, incidentally, demonstrate the level of MDA similar to that of the afore-mentioned study.

The influence of smoking on MDA in EBC is unclear. Bartoli et al. (2011) have demonstrated that although MDA is the highest in COPD patients, compared to patients with other chronic respiratory diseases, its value was similar in current smokers and ex-/non-smokers with COPD. On the other hand, smoking asthmatics have a higher MDA level than ex-/non-smoking asthmatics. The present finding of a lower level of TNF-α and higher of MDA in EBC after smoking in healthy smokers, compared to COPD patients, is difficult to interpret, due to the lack of relevant studies comparing both biomarkers after acute CS exposure.

There is a strong body of evidence that markers of oxidative stress respond rapidly to CS exposure. CS-induced cytokine alterations in various biologic materials require more time due to complex metabolic pathways, which lead to cytokine production. However, inflammatory cells have the potential of rapid cytokine release. Ryder et al. (2002) have shown that mononuclear blood cells release IL-1β and TNF-α as early as 2–5 min of *in vitro* smoke exposure. Clinical studies show that the EBC cytokine content may change within less than 60 min after CS exposure. Garey et al. (2004) have shown that TNF-α level tends to be higher than the baseline value, while IL-1β significantly decreases in EBC 30 min after smoking one cigarette by healthy smokers. Although we measured the cytokine levels after a twice longer time between smoking and EBC collection, we failed to show significant differences between the pre- and post CS values. One of the most important factors that might have been responsible for this situation is the probable lower tar and nicotine content in cigarettes used in our study. The study by Garey et al. (2004) was published over a decade ago, when the restrictions for the tobacco industry were milder; the participants of that study smoked a cigarette of their choice, and the tar and nicotine content were not mentioned in the study protocol. Durak et al. (2000) have investigated serum markers of oxidative stress in healthy subjects after smoking full flavor and full flavor low tar cigarettes (23 mg and 12 mg tar content, respectively) and demonstrated a lower burden of oxidative stress after smoking the latter. The cigarettes used in our study had 6 mg tar content. We may assume that lower tar and nicotine content induce smaller alterations in the cytokine profile of biological samples from smokers.

Interestingly, the present results showed a negative relationship between EBC TNF-α concentration after smoking and FEV_1/FVC in healthy smokers. To our knowledge, this correlation has not been reported before. This relationship suggests that monitoring of TNF-α level in smokers may be useful in the identification of subjects at risk for COPD. TNF-α is considered a major player in connective tissue damage in response to CS (Comandini et al. 2009), promoting COPD-related inflammation at the local and systemic levels (Górska et al. 2010; Tanni et al. 2010). The influence of smoking on annual FEV_1 decline has been well documented (Fletcher and Peto 1977), and a decreased FEV_1/FVC ratio is the main functional criterion for COPD diagnosis (GOLD 2015). Therefore, it seems that smoking young subjects who demonstrate a relationship between FEV_1/FVC and airway inflammation markers involved in the pathogenesis of COPD should be the target for smoking-cessation initiatives and COPD screening programs. That is supported by a review of Kotz et al. (2007) who provide evidence that smoking has early negative effects on lung function in young smokers with a short smoking history.

Our study has several limitations. Firstly, there was a significant difference between both analyzed groups in terms of smoking history. This was imposed by one of the aims of the study, i.e., the comparison of the effect of CS in young healthy smokers and patients with COPD. Healthy subjects obviously had to have a less relevant smoking history, mainly due to a shorter time of smoking. Secondly, our measurements were performed only at two time points, i.e., before and 60 min after smoking. Serial measurements could better reflect the airway inflammatory response to CS over time, but their application in the clinical setting is rather difficult due to time duration and cost. Our study aimed at the evaluation of a rapid airway inflammatory response to CS exposure, hence we chose the assessment at two time points.

5 Conclusions

Acute cigarette smoke exposure does not increase the concentration of TNF-α, IL-1β, and malondialdehyde in the exhaled breath condensate in both smoking COPD patients and healthy asymptomatic smokers. Short-term cigarette smoke-induced oxidative stress is higher in young smokers than in COPD patients, what indicates a greater susceptibility to cigarette smoke content of young smokers.

Acknowledgements The authors would like to thank Piotr Korczynski for his assistance in the preparation of the manuscript.

Conflicts of Interest The authors declare no conflict of interest related to the contents of this article.

References

Bartoli ML, Novelli F, Costa F, Malagrino L, Melosini L, Bacci E, Cianchetti S, Dente FL, Di Franco A, Vagaggini B, Paggiaro PL (2011) Malondialdehyde in exhaled breath condensate as a marker of oxidative stress in different pulmonary diseases. Mediators Inflamm. doi:10.1155/2011/891752

Comandini A, Rogliani P, Nunziata A, Cazzola M, Curradi G, Saltini C (2009) Biomarkers of lung damage associated with tobacco smoke in induced sputum. Respir Med 103:1592–1613

Corradi M, Massimo I, Andreoli R, Manini P, Caglieri A, Poli D, Alinovi R, Mutti A (2003) Aldehydes in exhaled breath condensate in patients with chronic obstructive pulmonary disease. Am J Respir Crit Care Med 167:1380–1386

Durak I, Bingöl NK, Avci A, Cimen MY, Karaca L, Ozturk HS (2000) Acute effects of smoking of cigarettes with different tar content on plasma oxidant/antioxidant status. Inhal Toxicol 12:641–647

Fletcher C, Peto R (1977) The natural history of chronic airflow limitation. Br Med J 1:1645–1648

Garey KW, Neuhauser MM, Robbins RA, Danziger LH, Rubinstein I (2004) Markers of inflammation in exhaled breath condensate of young healthy smokers. Chest 125:22–26

GOLD (2015) Global initiative for obstructive lung disease. Global strategy for the diagnosis, management and prevention of chronic obstructive pulmonary disease. 2015 update. http://www.goldcopd.org. Accessed 18 Jan 2016

Górska K, Maskey-Warzęchowska M, Krenke R (2010) Airway inflammation in chronic obstructive pulmonary disease. Curr Opin Pulm Med 16:89–96

Hoonhorst SJM, Timens W, Koenderman L, Lo Tam Loi AT, Lammers JW, Boezen HM, van Oosterhout AJ, Postma DS, Ten Hacken NH (2014) Increased activation of blood neutrophils after cigarette smoking in young individuals susceptible to COPD. Respir Res 15:121–130

Horvath I, Hunt J, Barnes PJ et al (2005) Exhaled breath condensate: methodological recommendations and unresolved questions. Eur Respir J 26:523–548

Koczulla AR, Noeske S, Herr C, Jörres RA, Römmelt H, Vogelmeier C, Bals R (2010) Acute and chronic effects of smoking on inflammation markers in exhaled breath condensate in current smokers. Respiration 79:61–67

Konstantinidi EM, Lappas AS, Tzortzi AS, Behrakis PK (2015) Exhaled breath condensate: technical and diagnostic aspects. ScientificWorldJournal 2015:435160. doi:10.1155/2015/435160

Kostikas K, Minas M, Nikolaou E et al (2013) Secondhand smoke exposure induces acutely airway acidification and oxidative stress. Respir Med 107:172–179

Kotz D, van de Kant K, Jöbsis Q, van Schayck CP (2007) Effects of tobacco exposure on lung health and pulmonary biomarkers in young, healthy smokers aged 12–25 years: a systematic review. Expert Rev Respir Med 1:403–418

Lommatzsch M, Bratke K, Knappe T, Bier A, Dreschler K, Kuepper M, Stoll P, Julius P, Virchow JC (2010) Acute effects of tobacco smoke on human airway dendritic cells in vivo. Eur Respir J 35:1130–1136

Miller MR, Hankinson J, Brusasco V et al (2005) Standardisation of spirometry. Eur Respir J 26:319–338

Papaioannou AI, Koutsokera A, Tanou K, Kiropoulos TS, Tsilioni I, Oikonomidi S, Liadaki K, Pournaras S, Gourgoulianis KI, Kostikas K (2010) The acute effect of smoking in healthy and asthmatic smokers. Eur J Clin Investig 40:103–109

Pellegrino R, Viegi G, Brusasco V et al (2005) Interpretative strategies for lung function tests. Eur Respir J 26:948–968

Rovina N, Koutsoukou A, Koulouris NG (2013) Inflammation and immune response in COPD: where do we stand? Mediators Inflamm 2013:413735. doi:10.1155/2013/413735

Ryder MI, Saghizadeh M, Ding Y, Nguyen N, Soskolne A (2002) Effects of tobacco smoke on the secretion of interleukin 1β, tumor necrosis factor α and transforming growth factor β from peripheral blood mononuclear cells. Oral Microbiol Immunol 17:331–336

Tanni SE, Pelegrino NRG, Angeleli AYO, Correa C, Godoy I (2010) Smoking status and tumor necrosis factor-alpha mediated systemic inflammation in COPD patients. J Inflamm 7:29–35

van der Toorn M, Slebos DJ, de Bruin HG, Gras R, Rezayat D, Jorge L, Sandra K, Oosterhout AJ (2013) Critical role of aldehydes in cigarette smoke-induced acute airway inflammation. Respir Res 14:45–56

van der Vaart H, Postma DS, Timens W, ten Hacken NH (2004) Acute effects of cigarette smoke on inflammation and oxidative stress: a review. Thorax 59:713–721

van der Vaart H, Postma DS, Timens W, Hylkema MN, Willemse BW, Boezen HM, Vonk JM, de Reus DM, Kauffman HF, ten Hacken NH (2005) Acute effects of cigarette smoking on inflammation in healthy intermittent smokers. Respir Res 6:22–32

WCRF (2016) World Cancer Research Fund International. Lung cancer statistics. http://www.wcrf.org/int/cancer-facts-figures. Accessed 18 Jan 2016

WHO (2016) Prevalence of tobacco use. Global health observatory data. http://www.who.int/gho/tobacco/use. Accessed 18 Jan 2016.

Advs Exp. Medicine, Biology - Neuroscience and Respiration (2017) 27: 81–85
DOI 10.1007/5584_2016_60

Published online: 29 July 2016

Cigarette Smoking and Respiratory System Diseases in Adolescents

Agnieszka Saracen

Abstract

Respiratory system diseases are common in youngsters, smoking being one of the main cause of them. In this article, results are presented of a survey-type study on smoking and respiratory malady conducted in 3108 high school students from the Mazovian Region in Poland. The questionnaire made for this study contained questions concerning the health status, chronic diseases, and the cigarette smoking habit. The subjects were high school student aged 15–19. Overall, 1694 males and 1414 females were enrolled in the study. Regarding males, 66.4 % of them were non-smokers, 18.1 % smoked up to 20 cigarettes daily, and 15.5 % smoked more than 20 cigarettes daily; 12.5 % of all smokers smoked longer than one year. Overall, 38.5 % of males reported symptoms of chronic bronchitis. When stratified by the smoking habit, chronic bronchitis was reported by 21 % of non-smokers and 71 % of all smokers. Regarding females, 77 % of them were non-smokers, 16 % smoked up to 20 cigarettes daily, and 7 % more than 20 cigarettes daily; 8 % of all smokers smoked longer than one year. Overall, 35 % females reported symptoms of chronic bronchitis. When stratified by the smoking habit, chronic bronchitis was reported by 23 % of non-smokers and 75 % of smokers. Bronchial asthma was reported by 22 (0.7 %) subjects, none of them was a smoker. In conclusion, males more often than females smoked cigarettes. The number of persons complaining of symptoms of chronic bronchitis was markedly higher in the group of smokers. The study shows that smoking is a key cause of chronic bronchitis in adolescents. That implies a need for enhanced educational activity on the adverse effects of smoking and undertaking active anti-smoking campaigns at the level of high school.

The original version of this chapter has been revised. An erratum to this chapter can be found at DOI 10.1007/978-3-319-44488-8_11

A. Saracen (✉)
Faculty of Health Sciences and Physical Education, Kazimierz Pulaski University of Technology and Humanities in Radom, 27 Chrobrego Street, 26-600 Radom, Poland
e-mail: saracena@op.pl

Keywords
Adolescents • Asthma • Chronic bronchitis • Cigarette smoking • Health care

1 Introduction

Cigarette smoking is one of the most important causes of different pulmonary symptoms and diseases, from cough up to malignant neoplasms (Beasley et al. 2015; Islami et al. 2015). In a significant portion of persons cigarette smoking begins early, between 10 and 18 years of age (Bielska et al. 2015; Mazur et al. 2008). The key role in a decision to start smoking plays the family and school environment (Zaborskis and Sirvyte 2015; Saracen 2010; Kowalewska and Mazur 2008). Young people who smoke typically reveal symptoms of chronic bronchitis (Chiarini et al. 2015; Saracen 2010). The goal of the present study was to conduct an epidemiological survey-type study concerning the presence of respiratory complaints and maladies in a cohort of Polish high school students who self-reported the cigarette smoking habit.

2 Methods

The study was approved by the Ethics Board for Human Research of the Kazimierz Pulaski University of Technology and Humanities in the city of Radom, Poland. Three thousand one hundred and eight students, aged 15–19 years, who represented 19 high school institutions of the Mazovian voivodship in Central Poland, were enrolled into the investigation. An author's questionnaire, specifically written for the investigation, included questions concerning demographic data such as gender, age, place of living (country, city), and family economic status, as well as the presence of symptoms of a pulmonary malady, information on medical diagnosis and treatment, cigarette smoking habit among respondents and their family members, the amount of cigarettes smoked daily, and the length of smoking period. The questionnaire consisted of 23 classification type items, with YES and NO responses. A self-reported questionnaire seems an effective and optimal method of gathering epidemiological data on smoking habit in young populations (Chiarini et al. 2015; Saracen 2010).

Categorical data were statistically examined for the significance of the association between corresponding structural classifications of the population groups studied with a chi-squared test or Fisher's exact test. Associations between variables were assessed with Pearson's correlation coefficient. A p-value of <0.05 was used to define statistical significance.

3 Results

Three thousand one hundred and eight adolescents enrolled into the study were stratified into two age-groups: 15–16 years consisting of 1430 (46 %) subjects and 17–19 years consisting of 1678 (54 %) subjects. Stratification of the whole cohort by gender was as follows: 1694 (55 %) males and 1414 (45 %) females. One thousand four hundred twenty eight (46 %) subjects lived in cities and 1680 (54 %) lived in the country.

Of the 1694 males, 1125 (66 %) were non-smokers and 569 (34 %) smoked cigarettes. Of the male smokers, 307 (54 %) smoked up to 20 cigarettes per day and 262 (46 %) smoked more than 20 cigarettes per day. Two hundred and twelve (37 %) smokers smoked cigarettes for longer than a year.

Overall, 640 male subjects reported chronic bronchitis, with the following distribution by the smoking status: 404 (71 %) subjects of the 569 smokers and 236 (21 %) subjects of the 1125 non-smokers. The prevalence of chronic bronchitis was significantly greater in smokers ($p = 0.001$).

Table 1 Cigarette smoking and prevalence of chronic bronchitis in adolescent smokers stratified by gender

	Males; n (%)		Females; n (%)	
	Smokers	Non-smokers	Smokers	Non-smokers
	569 (34)[a]	1125 (66)	325 (23)	1089 (77)
Cigarettes/day		–		–
< 20	307 (54)		226 (70)[b]	
> 20	262 (46)[c]		99 (30)	
Incidence of chronic bronchitis	404 (71)[d]	236 (21)	245 (75)[d, e]	250 (23)
Parental smoking				
neither parent	0	450 (40)	0	436 (40)
one parent	228 (40)	450 (40)	130 (40)	436 (40)
both parents	341 (60)[f]	225 (20)	195 (60)[f]	218 (20)

[a]males were smokers more often than females ($p < 0.001$)
[b]greater percentage of females than males smoked fewer than 20 cigarettes/day
[c]greater percentage of males than females smoked more than 20 cigarettes/day ($p < 0.001$)
[d]greater percentage of both male and female smokers presented symptoms of chronic bronchitis more often than the corresponding groups of non-smokers ($p < 0.001$)
[e]greater percentage of female than male smokers complained of chronic bronchitis ($p < 0.01$)
[f]greater percentage of male and female smokers had both smoking parents compared with non-smokers ($p < 0.001$)

Of the 1414 females, 1089 (77 %) were non-smokers and 325 (23 %) smoked cigarettes. Of the female smokers, 226 (70 %) smoked up to 20 cigarettes per day and 99 (30 %) smoked more than 20 cigarettes per day. One hundred and thirteen (35 %) female smokers smoked cigarettes for longer than a year. Overall, 495 female subjects reported chronic bronchitis, with the following distribution by the smoking status: 245 (76 %) subjects of the 325 smokers and 250 (23 %) subjects of the 1089 non-smokers. Akin to the male group with respiratory symptoms, the prevalence of chronic bronchitis was also significantly greater in female smokers ($p = 0.001$).

A comparison of the gender groups revealed that males were more frequent cigarette smokers ($p = 0.0001$) and more often were chain smokers ($p = 0.001$) than females. Despite that a greater number of males smoked and they smoked more cigarettes per day than females, a greater percentage of female smokers complained of chronic bronchitis than that of male smokers ($p = 0.01$), suggesting a greater vulnerability of female's respiratory tract to detrimental effects of smoking. The place of living, be it city or country, and economical status of the family had no association to the prevalence of chronic bronchitis in either smokers or non-smokers. Twenty two respondents (0.7 %) reported bronchial asthma. None of them was a cigarette smoker.

In 40 % of cases, both parents of non-smoking males and females did not smoke cigarettes. In contrast, there was no case of both non-smoking parents among smoking males and females. Either one (228; 40 %) or both parents (341; 60 %) smoked in the group of smoking males. The corresponding figures for female smokers' parents were 130 (40 %) – single parent and 195 (60 %) – both parents smoking. These differences were significant ($p = 0.001$). Detailed data on the prevalence of smoking and chronic bronchitis are summarized in Table 1.

When the younger 15–16 years of age and the older 17–19-years of age groups were compared, no appreciable differences were found in any of the study aspects above outlined (data not shown).

4 Discussion

Cigarette smoking is an important social and epidemiological problem (Brinker et al. 2015; Kandel et al. 2015; Kowalewska 2008). In young people, smoking is the main cause of chronic respiratory diseases, mainly bronchitis, but it also produces a number of significant health disturbances, leading to decreased fertility, increased insulin resistance, or thyroid insufficiency (Lingappa et al. 2015; Meral et al. 2015). Despite intensive campaigns against cigarette smoking, it remains a frequent type of life style among youngsters; in a sense recently having a boost in the form of e-cigarettes and smokeless tobacco that actually are conducive to the persistence of a smoking habit (Piñeiro et al. 2015; Wolfson et al. 2015). Our study shows that the anti-cigarette campaigns in Poland do not stop the youths and adolescents from smoking cigarettes. Further, percentage of young smokers have increased in the last 5 years by about 10 % (Kłos and Gromadecka-Sutkiewicz 2008; Kowalewska 2008; Kowalewska and Mazur 2008). Adolescents are savvy about healthy life style. Their attitude toward healthy behaviors is shaped by family and school environments (Niu et al. 2015; Saracen 2010). We also found in the present study that all smoking students had smoking parents, at least one of them. That points to the need of initiating more anti-smoking activities and measures directed toward parents and family environment. We also should take into account the probable disadvantageous influence on the number of smokers of e-cigarette promotion (Piñeiro et al. 2015). Our findings show that cigarette smoking starts off very early; about 40 % of school students had begun smoking before 15 years of age. The percentage of smokers in the first and last grade was similar, which shows that if there are quitters along the school years, they are substituted for by newcomers to smoking. In line with the literature data (Kandel et al. 2015; Piñeiro et al. 2015) we found that females smoked less often than males did, but the percentage of chronic bronchitis was higher among smoking females than males. Overall, more than 70 % of smoking students presented symptoms of, and were medically treated for, chronic bronchitis. This number points attention to the extent of likely detrimental health problems developing in later life and to the epidemiological and economical seriousness of the issue. In addition, an increasing cost of cigarettes seems to bear little effect on the number of smokers, particularly in relatively wealthy regions of the country, like the Mazovian voivodship in Poland, having not much of an effect on economic-driven restriction of adolescents' access to cigarettes. All students of the present study had unlimited access to electronic media, both at home and school. Despite constant information concerning harmful effects of cigarettes, our findings failed to unravel a reduction in tobacco usage among Polish teenagers and adolescents. There is an apparent necessity to develop new strategies to combat cigarette smoking and to reduce the number of smoking-related chronic pulmonary diseases.

5 Conclusions

This study demonstrates that despite intensive informative activities directed at sharing knowledge on the harmful effects of tobacco smoking, the number of smoking teenagers increases and more than 30 % of school students admit smoking cigarettes. Smoking begins at an early time of life, before 15 years of age, and more than 70 % of adolescent smokers reveal symptoms of chronic bronchitis. Males smoke cigarettes more often than females do, but females appear more vulnerable to lung detrimental effects of smoking. All young smokers have at least one smoking parent. The widespread incidence and severity of the adolescent smoking lead to a conclusion that anti-smoking measures should concentrate not only on adolescents, but also, and perhaps in the first place, on the smoking parents. Intensive anti-smoking activities should embrace school counselors, carers, and educators, whoever they may be.

Conflicts of Interest The author declares no conflict of interest in relation to his article.

References

Beasley R, Semprini A, Mitchell EA (2015) Risk factors for asthma: is prevention possible? Lancet 386:1075–1085

Bielska DE, Kurpas D, Nitsch-Osuch A, Gomółka E, Ołdak E, Chlabicz S, Owłasiuk A (2015) Exposure to environmental tobacco smoke and respiratory tract infections in pre-school children – a cross-sectional study in Poland. Ann Agric Environ Med 22:524–529

Brinker TJ, Stamm-Balderjahn S, Seeger W, Klingelhöfer D, Groneberg DA (2015) Education Against Tobacco (EAT): a quasi-experimental prospective evaluation of a multinational medical-student-delivered smoking prevention programme for secondary schools in Germany. BMJ Open 5:e008093. doi:10.1136/bmjopen-2015-008093

Chiarini M, Saulle R, Panaro AS, La Torre G (2015) Validation of a questionnaire to assess knowledge, attitudes and behaviors towards smoking among nursing students: a pilot study in Piedmont region. Prof Inferm 68:183–189

Islami F, Torre LA, Jemal A (2015) Global trends of lung cancer mortality and smoking prevalence. Transl Lung Cancer Res 4:327–338

Kandel DB, Griesler PC, Hu MC (2015) Intergenerational patterns of smoking and nicotine dependence among US adolescents. Am J Public Health 105:e63–e72. doi:10.2105/AJPH.2015.302775

Kłos J, Gromadecka-Sutkiewicz M (2008) Smoking as an aspect of lifestyle among 18-year-old secondary school students in Poznan. Przegl Lek 65:553–559

Kowalewska A (2008) The age of tobacco initiation and tobacco smoking frequency among 15 year-old adolescents in Poland. Przegl Lek 65:546–548

Kowalewska A, Mazur J (2008) Tobacco smoking by family members at home and adolescents' subjective health. Przegl Lek 65:549–552

Lingappa HA, Govindashetty AM, Puttaveerachary AK, Manchaiah S, Krishnamurthy A, Bashir S, Doddaiah N (2015) Evaluation of effect of cigarette smoking on vital seminal parameters which influence fertility. J Clin Diagn Res 9:EC13–EC15

Mazur J, Woynarowska B, Kowalewska A (2008) Selected indicators of tobacco smoking in 15-year-old students in Poland in relation to international statistics. Przegl Lek 65:541–545

Meral I, Arslan A, Him A, Arslan H (2015) Smoking-related alterations in serum levels of thyroid hormones and insulin in female and male students. Altern Ther Health Med 21:24–29

Niu L, Luo D, Silenzio VM, Xiao S, Tian Y (2015) Are informing knowledge and supportive attitude enough for tobacco control? A latent class analysis of cigarette smoking patterns among medical teachers in China. Int J Environ Res Public Health 25:12030–12042

Piñeiro B, Correa JB, Simmons VN, Harrell PT, Menzie NS, Unrod M, Meltzer LR, Brandon TH (2015) Gender differences in use and expectancies of e-cigarettes: online survey results. Addict Behav 14:91–97

Saracen A (2010) Health behaviors of high school students. Hygeia Public Health 45:70–73

Wolfson M, Suerken CK, Egan KL, Sutfin EL, Reboussin BA, Wagoner KG, Spangler J (2015) The role of smokeless tobacco use in smoking persistence among male college students. Am J Drug Alcohol Abuse 16:1–6

Zaborskis A, Sirvyte D (2015) Familial determinants of current smoking among adolescents of Lithuania: a cross-sectional survey 2014. BMC Public Health 15:889

ERRATUM

Advs Exp. Medicine, Biology - Neuroscience and Respiration (2017) 27: E1–E2
DOI 10.1007/978-3-319-44488-8_11

Respiratory Treatment and Prevention

Mieczyslaw Pokorski

Erratum to:
Advs Exp. Medicine, Biology - Neuroscience and Respiration
DOI 10.1007/978-3-319-44488-8

DOI 10.1007/978-3-319-44488-8_11

Due to an unfortunate production error the chapters related to this book were published by mistake multiple times.

As a consequence the DOI numbers of the old chapters had to be aliased to the new DOI numbers.

The DOI numbers of the removed duplicates and their related final and correct DOI numbers are listed below here for easy reference:

Final and correct DOI	Removed DOI duplicate	Removed DOI duplicate
10.1007/5584_2016_42	10.1007/978-3-319-44488-8_42	10.1007/5584_2016_138
10.1007/5584_2016_43	10.1007/978-3-319-44488-8_43	10.1007/5584_2016_141
10.1007/5584_2016_44	10.1007/978-3-319-44488-8_44	10.1007/5584_2016_142
10.1007/5584_2016_45	10.1007/978-3-319-44488-8_45	10.1007/5584_2016_143
10.1007/5584_2016_46	10.1007/978-3-319-44488-8_46	10.1007/5584_2016_145
10.1007/5584_2016_47	10.1007/978-3-319-44488-8_47	10.1007/5584_2016_146
10.1007/5584_2016_55	10.1007/978-3-319-44488-8_55	10.1007/5584_2016_137
10.1007/5584_2016_61	10.1007/978-3-319-44488-8_61	10.1007/5584_2016_139
10.1007/5584_2016_59	10.1007/978-3-319-44488-8_59	10.1007/5584_2016_140
10.1007/5584_2016_60	10.1007/978-3-319-44488-8_60	10.1007/5584_2016_144

The updated online version of the book can be found at
http://dx.doi.org/10.1007/978-3-319-44488-8

In addition the original online first copyright year of the chapters had been changed by accident at the first publication of this book. This has been corrected to the original online first copyright year.

The publishers apologies to the authors and readers for the inconvenience.

Advs Exp. Medicine, Biology - Neuroscience and Respiration (2017) 27: 87–88
DOI 10.1007/5584_2016

Index

A
Adolescents, 81–84
Air pollution, 10, 11, 14–16
Airway collapse, 65
Airway obstruction, 64
Antituberculotics, 22
Aspergillosis, 27–32
Aspergillus fumigatus, 29, 30, 32, 59
Asthma, 83

B
Bacteria, 20, 24, 48–51, 54, 59
Bronchoalveolar lavage fluid (BALF), 2–6, 28, 30, 74

C
Cardiopulmonary disease, 9–16
Cefotaxime, 48–50
Cerebrospinal fluid (CSF), 53–60
Chronic bronchitis, 42, 82–84
Cigarette smoking (CS), 73–79, 81–84
City, 11, 12, 14, 82, 83
CSF. *See* Cerebrospinal fluid (CSF)

D
Diagnosis/diagnostics, 3, 5, 19–24, 27–32, 36–39, 44, 45, 54, 55, 58, 60, 64, 66, 75, 79, 82
DNA, 31, 48–50, 53–60

E
ELISA. *See* Enzyme-linked immunosorbent assay (ELISA)
Encephalitis, 54, 58, 59
Enzyme-linked immunosorbent assay (ELISA), 3, 28, 29, 38, 39, 75

G
Galactomannan (GM) antigen, 28–32
Gentamicin, 48–51
Granulocytes, 48, 50, 51

H
Health care,

I
Infection, 10, 20–24, 28–32, 36–39, 42–44, 47, 48, 50, 54, 58–60, 75
Inflammation, 51, 74, 79

L
Life style, 64, 84
Lungs, 2, 3, 5, 6, 21–24, 28, 50, 75, 79, 84
  cancer, 9–16, 74

M
Malaria
  parasites, 36–39
  symptoms, 36–39, 41
  treatment, 36, 37, 43, 44
Malondialdehyde (MDA), 74–79
Mandibular advancement appliance (MAA), 63–69
Metagenomics analysis, 53–60
Mosquito, 36
Mycobacteriosis, 23

N
Neutrophil extracellular traps (NETs), 47–51
Next-generation sequencing (NGS), 53–60
Non-tuberculous mycobacteria (NTM), 19–24

O
Osteoprotegerin (OPG), 1–6

P
Particulate matter (PM), 10–16
Pathogens, 20, 22, 24, 48, 49, 54, 55, 58, 59
Phagocytosis, 51
Pollutant concentration, 11, 12, 15
Population attributable fraction, 11, 12
Premature death, 10
Prophylaxis, 40, 43, 44

R
Respiratory tract, 2, 5, 24, 28–31, 74, 83

S
Sarcoidosis, 1–6
Sleep apnea, 63–69
Sleep disordered breathing, 63
Smoking, 73–79, 81–84
Soluble receptor activator of nuclear factor-kappaB (sRANKL) ligand, 1–6

T
Traveler malaria, 35–45
Treatment effectiveness, 65
Tropical disease, 36
Tumor necrosis factor alpha (TNF-α), 21, 74–79

U
Upper airways, 64, 65, 68, 69

V
Viruses, 38, 39, 48, 53–60

The manufacturer's authorised representative in the EU is Springer Nature Customer Service Centre GmbH, Europaplatz 3, 69115 Heidelberg, Germany. If you have any concerns regarding our products, please contact ProductSafety@springernature.com

Printed and bound by CPI Group (UK) Ltd, Croydon, CR0 4YY
15/07/2026
02167654-0004